EVERYTHING YOU NEED TO KNOW ABOUT LYME DISEASE AND OTHER TICK-BORNE DISORDERS

Karen Vanderhoof-Forschner, B.S., M.B.A., C.P.C.U., C.L.U.

Foreword by
Willy Burgdorfer, Ph.D., M.D. (hon.)
Afterword by Senator Joseph I. Lieberman

John Wiley & Sons, Inc.

New York • Chichester • Weinheim • Brisbane • Singapore • Toronto

This text is printed on acid-free paper.

Published by John Wiley & Sons, Inc.

The information contained in this book is not intended to serve as a replacement for professional medical advice. Any use of the information in this book is at the reader's discretion. The author and publisher specifically disclaim any and all liability arising directly or indirectly from the use or application of any information contained in this book. A health care professional should be consulted regarding your specific situation.

Library of Congress Cataloging-in-Publication Data

Vanderhoof-Forschner, Karen.
Everything you need to know about lyme disease and other tick-borne disorders / Karen Vanderhoof-Forschner.
p. cm.
Includes bibliographical references and index.
ISBN 0-471-16061-X (pbk. : alk. paper)
1. Lyme disease—Popular works. 2. Tick-borne diseases—Popular works. I. Title.
RC155.5.V36 1997
616.9'2—dc21 97-10073

Printed in the United States of America

10 9 8 7 6 5 4 3 2 1

This book is dedicated to several people who have made a profound difference in my life:

Jamie, my son (1985–1991), who was born into a world ignorant about Lyme disease and suffered because of it.
Christy, my daughter (1993), who was born into a world enlightened about many tick-borne disorders and is healthy.

. . . and the people who played a pivotal role in making the difference:

Dr. Willy Burgdorfer
Senator Joe Lieberman
Congressman George Hochbrueckner
and
my lifetime love, Thomas E. Forschner

CONTENTS

APPENDICES

FOREWORD

Since ancient times, ticks have been considered "disgusting parasitic animals," associated with a wide variety of terrestrial and flying vertebrates and even with a few marine snakes and lizards. In temperate zones and tropical countries, ticks surpass all other arthropods in the number and variety of disease agents they transmit to humans and their domestic animals. While some species are host-specific and rarely feed on humans, others will attack and feed on any blood source including humans that enter their biotopes.

Because of the already small sizes of ticks, paricularly of the larval and nymphal stages, and also because of the painless attachment and feeding, ticks often go unnoticed until they had dropped, leaving an itching and more or less severe inflammation at the site of the bite. Undoubtedly hundreds of Lyme disease patients were or still are unaware of having been bitten by the small "freckle with legs," that is, the nymphal deer tick.

Well, Karen Forschner was one such patient. After clearing brush with her husband, Tom, she came down with symptoms suggesting Lyme disease. She had never heard of this illness and its association with ticks crawling in her backyard and on her cats and dogs. The illness, unfortunately, remained undiagnosed through Karen's pregnancy and affected her son, Jamie, who became ill the second day of his life.

Suffering daily vomiting, eye tremors, paralysis, and brain damage, Jamie was seen by dozens of physicians who subjected him to brain scans, dyes, probes, muscle biopsies, operations, and hospitals without leading to a proper diagnosis and effective treatment.

It was then that Karen, determined to find the answers for Jamie's problems, began consulting the medical literature on Lyme disease. She found sufficient similarities to suggest that her son did inherit her illness—an assumption that eventually was confirmed by laboratory tests.

The more Karen read about Lyme disease, the more she became frustrated by the medical services and the limited information about a disease that infects large segments of the American people especially in the northeastern and midwestern states. Thus, in January 1988, with the help of her husband, Tom, and her parents, and supported by a board of directors with experts from the medical, scientific, and public advocacy, she started the Lyme Borreliosis Foundation—a nonprofit organization devoted to prevention, education, and treatment of Lyme disease.

As president of the foundation, Karen has been the driving force in providing the general public and medical profession with information about the disease. In promoting the foundation's activities, she has appeared on talk shows and symposia including *20/20, Inside Edition, Home Show, CNN, NBC, ABC,* and *CBS News,* and has been mentioned in the *Boston Globe,* the *New York Times, Macleans, Reader's Digest,* and *Family Circle* ("Women Who Make a Difference").

With her recent book, *Everything You Need to Know about Lyme Disease and Other Tick-Borne Disorders,* Karen provides a compendium of knowledge she has accumulated during her search for answers to her and Jamie's struggles with Lyme disease. In an easily understandable language, she introduces the reader to the complex clinical aspects of Lyme disease and discusses the often controversial issues of diagnosis and treatment. She also presents a historical analysis of findings that led to the emergence of Lyme disease in Europe and the United States. A special chapter is devoted to brief reviews of other tick-borne diseases in this country, namely Rocky Mountain spotted fever, relapsing fever, tularemia, Colorado tick fever, tick paralysis, and the emerging ehrlichioses. The causative agents of these entities have been shown to occasionally occur in the same tick species that transmit the agent of Lyme disease.

Several chapters deal with the biology of tick vectors, tick control, personal protection against tick bites, and the safe removal of attached and feeding ticks.

I am very pleased to write the foreword for this outstanding book. *Everything You Need to Know about Lyme Disease and Other Tick-Borne Disorders* belongs on every family, business, and health care professional's bookshelf. The information is scientifically based and the scope provides easy-to-understand information for the novice as well as new information for the well-informed. Unique to this book are the *outstanding* appendices. They include a tremendously valuable historical timeline of scientific discoveries (with the only available references to the original published works), a detailed bibliography, lists of support groups, resources available around the world, and Internet addresses.

Lyme disease continues to be the most prevalent tick-borne illness in this country and will continue to affect thousands of people, young and old, every year. *Everything You Need to Know about Lyme Disease and Other Tick-Borne Disorders* certainly lives up to its title and should be read by anyone living in tick-infested areas.

—Willy Burgdorfer, Ph.D., M.D. (hon.)

PREFACE

MY STRUGGLE WITH LYME DISEASE

My husband, Tom, and I lost our son when he was just 6 years old. Jamie died from a disease the nation's medical and scientific communities thought he could not get. They were wrong.

In 1985, I was bitten by a tick while I was pregnant and soon became ill with a multitude of problems, including a rash, serious joint swelling, and intense pain. A physician diagnosed me with crippling arthritis, told me it was incurable, and predicted I would eventually live my life in a wheelchair. He, too, was wrong.

My family's struggle with Lyme disease is a tragedy, but it is also a story of courage and hope. Out of our personal nightmare emerged the Lyme Disease Foundation and nationwide awareness of the risks of tick-borne illnesses.

Jamie was our first child, the blonde-haired, blue-eyed light of our life. But by the time he was 6 weeks old, he had begun to vomit repeatedly and had alarming eye tremors, a sign of brain infection. At 6 months, he showed signs of brain damage and malnourishment because he was unable to absorb his food. Hearing tests showed that he was totally deaf, although we later discovered that was not true. Physicians saw signs of permanent damage in his eyes. Surgery to realign his stomach to stop the life-threatening vomiting was unsuccessful.

Soon after Jamie's birth, when our struggles through one medical crisis after another were just beginning, a mysterious condition called Lyme disease was mentioned as a possible cause of the illness that had plagued me during my pregnancy. I was told that if that were so, a couple of antibiotic pills were all that was necessary to cure me. I was also assured that my infection could not have spread to the child I was carrying. Still, my doctor urged that we both be tested for Lyme. When laboratory tests were positive for both Jamie and I, we suddenly had some hope. Perhaps a simple antibiotic treatment would curb the vicious onslaught of symptoms.

Over the year, I learned there was nothing simple about treatment, but after repeated doses of antibiotics, I slowly regained my health. For Jamie, it was not to be. At first, he responded to the antibiotics but within weeks he relapsed. Local doctors repeatedly resisted our pleas to treat Jamie with long-term antibiotics, warning that the medication itself was dangerous. Tom and I were persistent, and intensive treatment temporarily improved my son's condition. For a time, he was able to attend kindergarten and began learning to talk. His muscle tone and vision improved and he was able to eat again. We were finally finding the little boy inside the ravaged body.

Sadly, it did not last. In his final weeks of life, Jamie's brain became inflamed and he began to have seizures. Our son was put on life support systems but the swelling could not be controlled. On June 21, 1991, he passed away.

It was the end of one struggle, but another one continued. In the years that we were fighting for Jamie's life, Tom and I had also given birth to an organization so that others would not have to face hardships like ours. We had been astonished at how little was known about the condition that eventually took Jamie from us. Most scientists did not believe it was possible to transmit Lyme infection in utero, but my own medical search demonstrated otherwise. The breadth of symptoms associated with the illness were poorly understood in this country and no one knew what to do with patients who did not respond to a single course of antibiotics.

Our personal plight had gained the attention of the television program *20/20,* and countless articles had been written about us in the national and international media. We were the catalyst for making Lyme disease a household term. As our story became known, our telephone began to ring. Hundreds of people around the country were recognizing their own symptoms in the descriptions of Jamie and me and they wanted to know where to go for help. The response from the scientific community was also deafening—researchers studying the disease in isolation from one another were desperate for opportunities to meet with colleagues to talk about the strange tick-borne diseases they were beginning to see. It had become obvious that an umbrella organization was badly needed to find the truth about Lyme disease.

To meet the needs of patients and researchers, we launched the Lyme Borreliosis Foundation in 1988, the first organization in the world dedicated exclusively to Lyme disease. Three years later, the name was changed to the Lyme Disease Foundation because it was so much easier to pronounce. I walked the halls of Congress, telling my family's story, educating representatives about the disease, and explaining why millions of dollars were needed to learn more about it. Senator Joseph Lieberman and Congressman George Hochbrueckner became invaluable partners. Thousands of patients and their physicians contacted their representatives to plead for funding. Together, our efforts were instrumental in securing the first targeted federal funding for Lyme disease education and research. The Foundation began to sponsor an annual, international scientific conference; made connections with European researchers, where this disease has been studied for more than a century; and for the first time established a network of community educators, support groups, and researchers.

Through our efforts, and that of other partnerships formed between concerned families and scientific professionals, attitudes toward Lyme disease have begun to change. Today, the medical community is finally addressing some neglected issues. They have awakened to the fact that the bacteria that causes Lyme disease can, under rare circumstances, be transmitted from mother to fetus via the placenta. Patients who do not respond to a single course of antibiotics are now offered other treatment options, although we still cannot predict what will work. And attention is finally being paid to prevention strategies that can keep us all safe.

We are a long way from eliminating Lyme disease and other tick-borne infections altogether, but there is no doubting our enormous progress. Armed with the information I have provided in this book, it is now possible to keep your family safe. And I expect the news will keep getting better as science advances and more people become knowledgeable about strategies for personal protection.

To think, it all began with the birth of a beautiful little boy named Jamie.

—Karen Vanderhoof-Forschner

ACKNOWLEDGMENTS

There are many people who have directly or indirectly helped to make this book possible. I am taking this opportunity to acknowledge their efforts.

My admiration goes to Thomas E. Forschner for his encouragement and unwavering belief that I could write this book. My gratitude goes to my parents Ruth and Irwin Vanderhoof for their unfailing belief that my volunteer effort at the Lyme Disease Foundation (LDF) is important and for being an inspiration in my life.

My appreciation goes to my agent, Ivy Fischer Stone, and my editor, PJ Dempsey, who stuck with this to a completed book. My special gratitude to Karyn Feiden, who took a dry, technical manuscript and made it more interesting and readable.

I am indebted to Willy Burgdorfer, Ph.D., M.D. (hon.), Scientist Emeritus of the National Institutes of Health, Senator Joseph I. Lieberman and family, Congressman George Hochbrueckner and family, and Berkley Bedell and family for their exemplary leadership in the Lyme disease movement.

I thank the following for their input, submission of material, and/or review of this book:

Satyen Banerjee, M.D., British Columbia Centre for Disease Control, Canada
Alan Barbour, M.D., University of California at Irvine
Edward Bosler, Ph.D., State University of New York at Stony Brook
Sandra Bushmich, D.V.M., University of Connecticut
Patricia Coyle, M.D., State University of New York at Stony Brook
Sam Donta, M.D., Chief of Infectious Diseases, Boston University
Sandra Evans, Lyme researcher and published author
Thomas E. Forschner, C.P.A., M.B.A., Executive Director, Lyme Disease Foundation
Duane Gubler, Ph.D., Centers for Disease Control and Prevention
Nick Harris, Ph.D., Founder, IGeneX
Bernie Hudson, M.D., Royal North Shore Hospital, Australia
Dave Kazarian, R.Ph., Infuserve America
Diane Kindree, B.S.N., Director, British Columbia Lyme Borreliosis Society, Canada
Ken Liegner, M.D., New York Medical College
James Miller, Ph.D., University of California at Los Angeles, School of Medicine

Lloyd Miller, D.V.M., New York
Terry Moore, TAGS, Australia
Gary Mount, United States Department of Agriculture
Julie Rawlings, M.P.H., Texas Department of Health
Carl Schreck, United States Department of Agriculture
Rudy Scrimenti, M.D., Medical College of Wisconsin
Dick Tilton, M.D., Boston Biomedica/North American Laboratories, Editor of the *Journal of Clinical Microbiology*
Gloria Wenk, education activist in New York

Overall thanks goes to members of Congress and their staffs, who helped support Lyme disease-related activities. A special thanks goes to Senator Alfonse D'Amato, Senator Bill Bradley, Senator Tom Harkins, Senator Edward Kennedy, Senator Frank Lautenberg, Senator Paul Simon, Congresswoman Rosa DeLauro, Congressman Rodney Frelinghuysen, Congressman Benjamin Gilman, Congressman Sam Gjedenson, Congresswoman Nancy Johnson, Congresswoman Barbara Kennelly, Congresswoman Nita Lowey, Congresswoman Marge Roukema, Congressman Jim Saxton, Congressman Chris Smith, Congressman Esteban Torres, and Congressman Bruce Vento.

I express my gratitude to the following people, the Board of Directors of the Lyme Disease Foundation, the LDF Scientific Advisors (more than 100 researchers, physicians, nurses, and other health care professionals), the LDF-New Jersey Task Force, and the many supporters of the Lyme Disease Foundation who helped move us all closer to answers to tick-borne disorders. I appreciate the help in typing by Cheryl Gillespie and Charley Vandergrift.

I appreciate the words of inspiration by Michael Klepper and Mary Tavon.

My admiration goes to governmental employees working on Lyme disease issues.

- The armed forces for being the first to offer real help to the LDF, developing better tick repellents and sharing educational information; once again they have demonstrated that the armed forces also protect the public through non-combat activities
- The Food and Drug Administration and United States Department of Agriculture for testing and monitoring Lyme-related products
- The Centers for Disease Control and Prevention and the National Institutes of Health for their ongoing search for answers and support of educational and research grants. I have always been especially impressed by the people at NIH's Rocky Mountain Laboratories for their ability to work with many divergent groups.

I would like to recognize those companies that have made a positive difference in Lyme disease:

Bill Perlberg and Mark Johnson, of Hartz Mountain, Inc., whose research grants help make possible development of vaccines, awareness of Lyme and Lyme-like pathogens in the Midwest, and an understanding of how the bacteria may do damage to the host. They also funded an educational program that played a major role in the country's level of awareness of Lyme disease and how to prevent the infection in people and pets. I wish I had known about tick-spread diseases so I could have used their products to help protect my pets.

The maker of Cutter Insect Repellent, an important company that has been honorable in their joint efforts with the LDF.

Family Circle and L'Oreal, for playing a major role in public education about this disease, which helped people learn prevention, early detection, and about the LDF.

Lastly, I express my appreciation to the people who produce Lyme disease products and services that will help make our lives better.

1

A PUBLIC HEALTH THREAT EMERGES

Remember the story of three people with blindfolds describing their first encounter with an elephant? One feels the long trunk and says with assurance, "Elephants are like a snake." The next puts his arms around the leg and says, "Elephants are like a trunk of a tree." And the third reaches up to the animal's huge body and announces, "Elephants are like a wall."

None of them is wrong. But each is presenting a picture that is incomplete and enormously misleading. Unless all three begin talking to each other and explaining why they reached their separate conclusions, the information is virtually useless to any effort to understand exactly what an elephant looks like.

That about sums up the struggles in Lyme disease. People have approached the problem from very different angles, and there has been too little effort at interdisciplinary dialogue. Dermatologists have studied the disorder because it is associated with rashes and other skin problems. Rheumatologists have viewed Lyme disease primarily as a cause of arthritic complaints, such as joint swelling and pain, and ophthalmologists have become involved with eye problems caused by the disease. Neurologists have been intrigued by the extent of brain involvement; cardiologists have been summoned when heart problems manifested themselves. But until recently, few of the experts were interacting with each other. The results have been delays in getting out the word about appropriate diagnosis and treatment, battles over scientific turf, and a painful fight for limited research dollars.

Fortunately, some collaborative efforts are finally taking place, and I am hopeful that this situation is beginning to turn around. It is high time. Lyme disease is the most common tick-borne infection in the United States and one of the fastest growing infectious diseases in North America. That makes it a public health epidemic of major proportions. According to the Centers for

Disease Control and Prevention (CDC), the branch of the federal government that monitors disease trends in the United States, more cases of Lyme disease were reported in 1994 than the *combined* total cases reported of measles, mumps, whooping cough, Rocky Mountain spotted fever, rubella, tularemia, diphtheria, cholera, brucellosis, malaria, leptospirosis, leprosy, encephalitis, plague, tetanus, trichinosis, typhoid fever, and Legionnaire's disease.

Nor is Lyme disease confined to this continent. In fact, it can be found on every continent in the world. Infection of both people and animals is worldwide. Infection has been cultured from ticks and humans in Europe, North America, Asia, and Australia. Cases have also been reported in South America and Africa. Researchers have even located infected ticks that help spread disease on seagulls and albatrosses in the Arctic Ocean and in Antarctica.

For some people, Lyme disease is a minor illness that is quickly diagnosed and easily cured. But for others, it becomes a personal nightmare, resulting in lost time from work or school, mental anguish, and sometimes permanent physical disability. In rare cases, it can be life-threatening. Virtually everyone in the United States is at risk for contracting Lyme disease or some other tick-borne infection until the outdoor temperature has dropped well below freezing for several consecutive days. The illness can also strike household pets and some livestock.

The elderly are at highest risk (for a profile of Lyme disease by age group, see Figure 1–1). Your chances of contracting the disease are higher if you spend a lot of time outdoors, especially if your work or recreational activities take you into forested areas. For example, two separate studies in New York and New Jersey found that outdoor workers are between two and four times as likely to contract Lyme disease as others. Forestry workers in the Netherlands and England are five times more likely than the general population to contract Lyme disease. Hunters and hikers face special danger. Owning a pet, especially a cat, also puts you at increased risk. Recently, studies have shown that even urban dwellers who use city parks may become infected with Lyme disease.

The disease is a costly burden on the health insurance system and a drain on employers who are losing their valued workers. In research published in the January/February 1993 issue of the actuarial journal *Contingencies,* which I coauthored with my father, Irwin T. Vanderhoof, Ph.D., a professor of economics and actuarial science, we were able to document that the cost of Lyme disease to society is a stunning $1 billion a year. Most of that comes from unnecessary or inappropriate medical care, lost productivity, legal fees, and other direct and indirect costs.

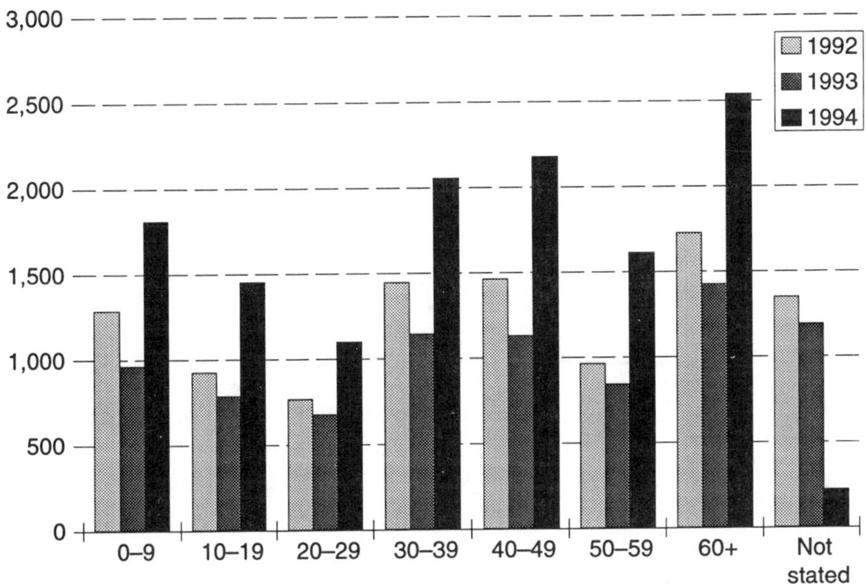

Summary of reported cases, grouped by 10-year age increments

	0–9	*10–19*	*20–29*	*30–39*	*40–49*	*50–59*	*60+*	*Not Stated*	*Total*
1992	1,249	945	825	1,428	1,453	969	1,694	1,332	9,895
1993	968	815	711	1,161	1,169	845	1,391	1,197	8,257
1994	1,819	1,463	1,113	2,062	2,207	1,612	2,540	227	13,043

Source: *Morbidity and Mortality Weekly Report,* U.S. Department of Health & Human Services, Public Health Service, Centers for Disease Control and Prevention.

Figure 1–1. Reported cases grouped by 10-year increments. Reprinted by permission of Karen Vanderhoof-Forschner.

Despite the human toll, much about Lyme disease remains uncertain and controversial. The research establishment believes that the current diagnostic tests are perfect. Tell that to patients who have been misdiagnosed with multiple sclerosis, Guillain-Barré syndrome, Alzheimer's disease, Lou Gehrig's disease, dementia, fibromyalgia, syphilis, or chronic fatigue syndrome. Some patients have even been improperly diagnosed with leukemia or other cancers. And let academic physicians explain why Lyme disease patients must see an average of five physicians before someone tells them what is causing their illness.

Some doctors still claim that a simple antibiotic regimen cures everyone.

But other physicians whose ailing patients have evidence of persisting infection know that is not so. Enormous gaps exist in knowledge about the risks of fetal transmission, the roots of symptoms that resemble multiple sclerosis and Alzheimer's disease, and the occasional instances when Lyme disease has led to death.

Clearly, there is work to be done.

UNDERSTANDING LYME DISEASE

The multiple symptoms of Lyme disease are caused by the *Borrelia burgdorferi (Bb)* spirochete, a type of bacteria characterized by a thin, spiral structure (Figure 1–2). By 1996, more than 100 diverse bacterial strains had been identified in the United States alone. Worldwide, 300 strains have been found, three times the number recognized when the Lyme Disease Foundation was founded in 1988. While some of the strains are quite similar, others, even within the same geographic region, are remarkably heterogenous.

The *Bb* spirochete is a parasite that is sustained in nature in the bodies of wild animals and is transmitted from one animal to another through the bite of an infected tick. In the United States, two members of the *Ixodes* family of ticks—popularly known as the black-legged tick *(Ixodes scapularis*) and the Western black-legged tick *(Ixodes pacificus*)—are by far the most common and well-established carriers of concern to humans. The lone star tick *(Amblyomma americanum*) can also transmit the Lyme disease–causing bacterium in this country, and there are reports that on rare occasions, the wood rat tick *(Ixodes neotomae*), and the rabbit tick *(Haemaphysalis leporispalustris*) have transmitted the infection to humans. The sheep tick *(Ixodes ricinus*), the hedgehog tick *(Ixodes hexagonus*), and the Taiga tick *(Ixodes persulcatus*) have been linked to disease in Europe and Asia. In Australia, the *Ixodes holocyclus* has been implicated. In the Arctic and Antarctic circles, the *Ixodes uriae* ticks are the culprits.

The Course of Lyme Disease

Lyme disease has been called the new "Great Imitator"—syphilis was first to earn that name—because of its ability to mimic other illnesses. It begins with an often-painless bite of a tick infected with *Bb*. Most people do not even notice that they have been bitten. As with other spirochetal infections, such as syphilis and relapsing fever, the first symptom of disease is often a local

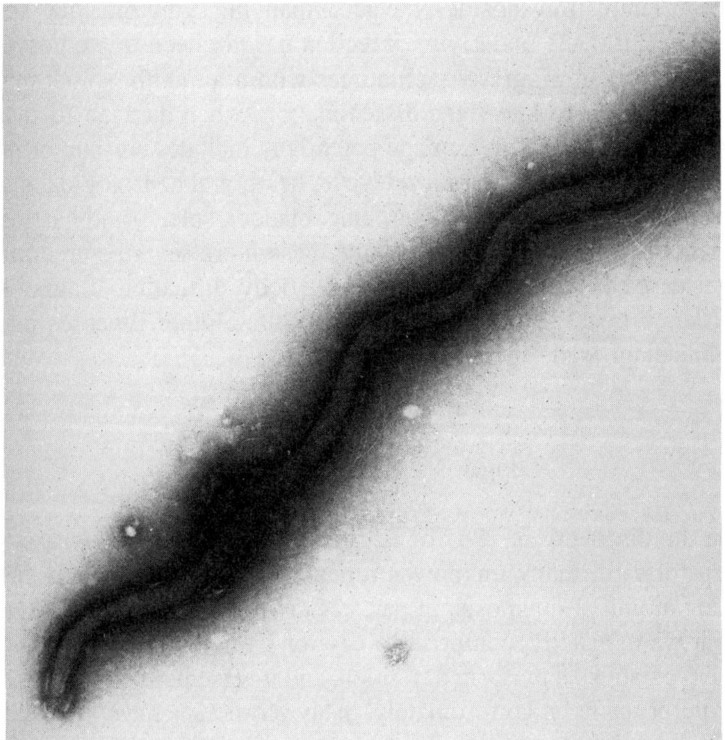

Figure 1–2. *Borrelia burgdorferi.* Reprinted by permission of Karen Vanderhoof-Forschner.

skin rash. The Lyme disease rash is called erythema migrans (EM), but it is not always present, and it, too, can be overlooked.

Contrary to some early scientific thinking, infection begins to be disseminated through the bloodstream within days of a tick bite, sometimes even before the rash appears. For example, Dr. Patricia Coyle, a neurologist at the Stony Brook campus of the State University of New York, has demonstrated that 67% of Lyme patients with an early rash also had *Bb* DNA in their spinal fluid, indicating that neurologic infection occurs very early in the disease.

Early dissemination is characterized by a period of inflammation and flu-like symptoms, such as headache, stiff neck, mild fever or chills, swollen glands, muscle aches, and fatigue. If the symptoms accompanying inflammation are severe, they can be an indication that the person will develop difficult-to-treat or recurring illness in the future. Often, however, the symptoms are not unduly alarming, and many people do not seek medical attention.

As the inflammation subsides, the accompanying symptoms may resolve on their own. But if the underlying infection has not been treated appropriately, the disease can progress—sometimes within a month, sometimes not for a year or more—to late-stage dissemination, when damage to multiple body systems can occur. This damage potentially includes chronic or relapsing problems of the skin, joint, muscle, eye, heart, and neurological system. Less often, some combination of liver, lung, bladder, spleen, and gastrointestinal ailments is also present. By the time these more severe symptoms appear, the disease can be securely lodged in the body, difficult to diagnose, and much harder to treat. Later symptoms also imitate other illnesses and can easily be misdiagnosed.

Diagnosis and Treatment

While the discovery in 1981 of *Bb* as the cause of Lyme disease was a major step forward, many unknowns remain. Most people have a difficult time getting an initial diagnosis, in part because of the substantial disagreement about what, exactly, comprises a case of Lyme disease. Only in 1990 did the CDC establish a uniform definition, and it was intended to be used for reporting purposes only. Unfortunately, many physicians have misused it as a diagnostic tool. It is sometimes the deciding factor for insurance reimbursement, as well. The problem with the CDC definition is that it is based solely on a few of the wide range of physical signs that a physician can see and measure and completely excludes all symptoms reported by the patient, such as pain, fatigue, or confusion.

According to the CDC's official definition (see page 186), a patient displaying the characteristic EM rash can definitively be considered to have Lyme disease. Alternatively, patients may be considered to have Lyme disease if they have at least one from a list of signs affecting the musculoskeletal, nervous, or cardiovascular systems, *and* if a laboratory is able to isolate the *Bb* spirochete, or detect a high level of antibodies to it, in a sample of blood or spinal fluid. Antibodies are the proteins produced by your immune system to fight off foreign substances such as toxins or infectious agents. While this narrow definition creates a research standard, it almost surely excludes many people who actually have the disease.

Why Lyme disease causes such a range of responses in those it strikes is also a mystery. How sick a person becomes may depend partly on exactly what infectious agents were injected into his or her body by the tick. Several researchers have recently shown that one tick can transmit several different strains of *Bb* at once. For example, in a study of ticks on Shelter Island, New York, published in the March 1996 issue of *Journal of Clinical Microbiology,* D. Guttman showed that 60 percent of the black-legged ticks studied were infected with more than one strain of *Bb.* Researchers in Belgium and Paris have also detected multiple strains of bacteria in either ticks or the individuals they bit. Ticks may also transmit more than one disease-causing infectious agent, known as a pathogen, with a cumulative effect that can be devastating. In the state of Connecticut, as many as 20 percent of all patients diagnosed with Lyme disease have also been diagnosed with at least one other tick-borne disease.

Some individuals who appear symptom-free may actually have a low level of *Bb* lingering in their bodies. There is no sure way to know whether or when a dormant infection will become active again. Other individuals truly are bacteria-free, but their anxiety about a relapse may tempt them to try potentially dangerous, expensive, and unnecessary treatments. Because existing laboratory tests cannot prove whether infection continues to linger in the body or is completely eliminated, it is impossible to say with assurance whether or not someone is completely cured. As a result, many patients remain in limbo, ever watchful for a new symptom that could strike at any time.

And finally, an increasing number of people who contract Lyme disease appear to have persisting infection. Chronic symptoms may linger or recur for months or years, and permanent damage can occur.

Despite the wealth of scientific information to the contrary, some academic researchers and the media are still sending a message that Lyme disease is always easy to cure. As a result, some people do not feel the need to take proper precautions against tick bites. Sadly, I am afraid it will take many more infections and a swelling population with long-term medical problems before an accurate message about the real dangers of disease—and the importance of prevention—gets widely circulated.

THE PROBLEM OF UNDERREPORTING

The CDC have accepted approximately 81,000 confirmed reports of Lyme disease since 1982. However, most of us working on the front lines of the epidemic believe this to be an enormous undercount. In our actuarial research, we were able to determine that only about 10% of qualifying cases are ever reported to the CDC. Adjusting the CDC figures pushes the case count to 810,000. Even that figure may be low. In 1996, the Connecticut Department of Public Health conducted a telephone survey that showed actual Lyme disease cases to be 13 times the number that were being reported. Reliable data about other tick-borne disorders is even sketchier, especially since Rocky Mountain spotted fever is the only other one for which the CDC requires reports.

To gauge whether even my estimate of 810,000 Lyme disease cases is accurate, look at Table 1, which provides the state-by-state CDC figures and then adjusts them tenfold to account for known underreporting. In New Jersey, the numbers suggest you would have to meet 93 people to find 1 person with CDC-defined Lyme disease. In California, the figure jumps to 1,523 people. If your own experience suggests that you are likely to meet far more than, say, 1 in 93 people with Lyme disease in New Jersey, as mine certainly does, then underreporting is even more of a problem than my own data suggest. I would not be surprised if at least one million Americans have contracted Lyme disease over the past 15 years. And since *Bb* has been in the United States for more than a century, the total number of infected people is likely to be far higher.

There are many reasons why the CDC's numbers are so low. For one, a formal case definition for Lyme disease was only established in 1990, physicians have only been required to report cases since 1994, and not until 1996 were all states actually in compliance with this requirement. Even now, CDC accepts only a narrow category of "confirmed" cases of Lyme disease, while

Table 1 Lyme Cases by State

State	*Population*	*Cases of Lyme disease reported since 1980**	*Adjusted cases meeting CDC criteria*	*How many people you would have to meet to find one Lyme patient*
AK	606,000	1	10	60,600
AL	4,219,000	109	1,090	3,871
AR	2,453,000	135	1,350	1,817
AZ	4,075,000	2	20	203,750
CA	31,431,000	2,064	20,640	1,523
CO	3,656,000	6	60	60,933
CT	3,275,000	11,321	113,210	29
DC	570,000	26	260	2,192
DE	706,000	658	6,580	107
FL	13,953,000	140	1,400	9,966
GA	7,055,000	1,177	11,770	599
HI	1,179,000	6	60	19,650
IA	2,829,000	175	1,750	1,617
ID	1,133,000	53	530	2,138
IL	11,752,000	241	2,410	4,876
IN	5,752,000	127	1,270	4,529

Table 1 Lyme Cases by State (continued)

State	*Population*	*Cases of Lyme disease reported since 1980**	*Adjusted cases meeting CDC criteria*	*How many people you would have to meet to find one Lyme patient*
KS	2,554,000	181	1,810	1,411
KY	3,827,000	170	1,700	2,251
LA	4,315,000	34	340	12,691
MA	6,041,000	1,831	18,310	330
MD	5,006,000	1,877	18,770	267
ME	1,240,000	113	1,130	1,097
MI	9,496,000	532	5,320	1,785
MN	4,567,000	1,478	14,780	309
MO	5,278,000	921	9,210	573
MS	2,669,000	34	340	7,850
MT	856,000	–	–	–
NC	7,070,000	604	6,040	1,171
ND	638,000	22	220	2,900
NE	1,623,000	46	460	3,528
NH	1,137,000	179	1,790	635
NJ	7,904,000	8,474	84,740	93
NM	1,654,000	15	150	11,027
NV	1,457,000	46	460	3,167
NY	18,169,000	30,930	309,300	59
OH	11,102,000	536	5,360	2,071
OK	3,258,000	218	2,180	1,494
OR	3,086,000	111	1,110	2,780
PA	12,052,000	7,558	75,580	159
RI	997,000	2,360	23,600	42
SC	3,664,000	90	900	4,071
SD	721,000	11	110	6,555
TN	5,175,000	223	2,230	2,321
TX	18,378,000	615	6,150	2,988
UT	1,908,000	24	240	7,950
VA	6,552,000	849	8,490	772
VT	580,000	66	660	879
WA	5,343,000	120	1,200	4,453
WI	5,082,000	4,087	40,870	124
WV	1,822,000	194	1,940	939
WY	476,000	43	430	1,107
Totals	260,341,000	80,833	808,330	

*Approximate, as the numbers continue to change

allowing some other tick-borne diseases, including Rocky Mountain spotted fever, to have additional categories, such as "probable" cases, using more relaxed criteria. CDC's decision not to use one of its other reporting categories, which include "clinically compatible" and "epidemiologically linked," omits other people who are likely to have Lyme disease.

Another cause of undercounting is the restrictive CDC definition. When the CDC's case definition became official, the Connecticut Department of Health reported a 15% drop in qualifying cases. And when the definition was narrowed somewhat in 1995, qualifying cases dropped even further and tended to exclude patients with later stages of disease. The New York State Department of Health observed a 31% drop in qualifying cases, with the great majority (81%) dropped being cases of late, disseminated disease.

Based on these statistical shifts, the media began to speculate that Lyme disease was less of a problem than had been predicted. Along with contributing to misleading statistics, a narrow case description means that research is targeted only at a portion of the disease spectrum, and that many ailing people are overlooked. Undercounting a disease also results in insufficient funding for both scientific research and public education. Until this is changed, scientists may never recognize the full scope of the Lyme disease problem.

There has also been a reporting bias that works against the nonwhite population. The overwhelming number of official Lyme disease cases have occurred in Caucasians. Most likely, many cases in people of color are not being properly diagnosed or treated. One reason is that the official definition refers to a red rash, but on dark skin, the EM rash looks more like a bruise and may not be noticed. Limited access to health care and the tendency of the media to refer to the disease as a "yuppie" illness may also contribute to the neglect of the minority community. The Lyme Disease Foundation has made a special effort to reach underserved populations, providing material in Spanish as well as captioned videotapes for the hearing impaired, because we know that Lyme is a disease that knows no boundaries.

THE SCIENCE OF LYME DISEASE

The key to prevention, the hope for a vaccine, improvements in diagnostic techniques, and more reliable treatment all depend on expanding our knowledge about *Bb*. Only by dedicating more research dollars to carefully targeted projects in basic research can we expect to make headway. Every discovery has the potential to advance the march of scientific progress.

Despite all that we know, countless questions about the Lyme disease bacterium remain unanswered, including these:

- Where does the *Bb* pathogen go in the body?
- What nourishment does *Bb* need to survive?
- How does *Bb* cause disease?
- Why does *Bb* appear to target nervous system tissue with special force?
- Is there a piece of the bacterium that can be used to stimulate a protective immune system reaction against all bacterial strains?
- Why do a few Lyme bacteria cause such a dramatic inflammatory response?

To visualize the *Bb* spirochete (Figure 1–3), imagine a long, thin snake wrapped lengthwise, as few as three times and as many as thirty, by hollow, thin, wire coils, known as flagella. Enclose this with two layers resembling cellophane, known as the outer membrane. Coat the entire package with a petroleum jelly-like substance that is the mucoid layer. Voila! This bacterium is many thousands of times larger than a virus, but a powerful microscope is still required to see it (Figure 1–4). Roughly 1,500 *Bb* must be laid end to end to equal 1 inch in length. Or, viewed from another perspective, it takes 100,000 spirochetes, laid side by side, to fill 1 inch of space.

The rigid, curly flagella give spirochetes their firm spiral shape and determine their serpentine, undulating motion. This motion may provide a vital

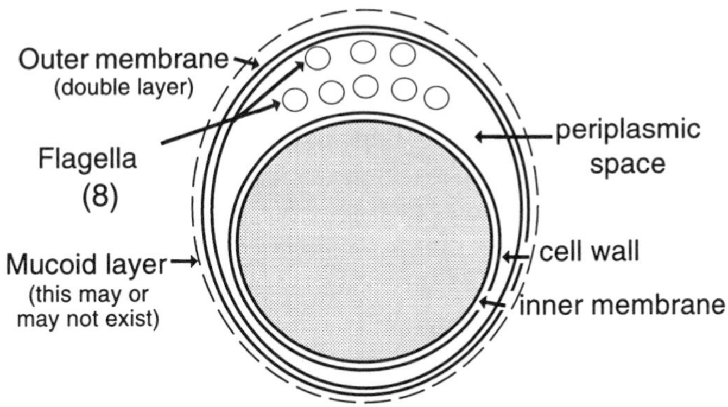

Figure 1–3. *Borrelia burgdorferi* in cross section. Copyright © Karen Vanderhoof-Forschner. Reprinted by permission of Karen Vanderhoof-Forschner.

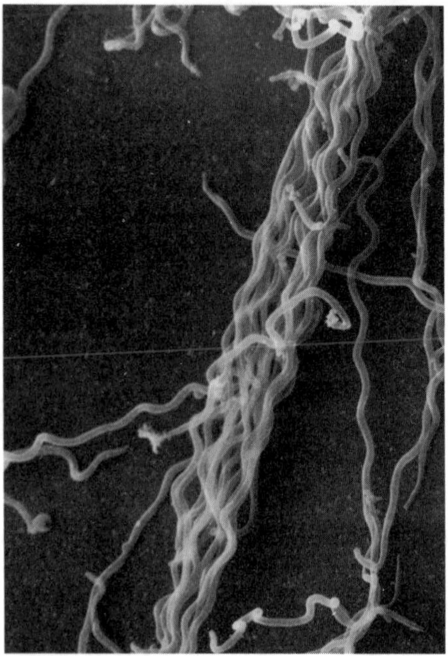

Figure 1–4. A culture of the Lyme bacterium.

clue as to how *Bb* invades human cells. A single Lyme disease bacterium may have between seven and thirty flagella, depending on the strain, but we don't know why the differences occur or what they imply about the severity of disease, if anything. Whether the strength of the human antibody response varies according to the number of flagella present also remains uncertain.

Another area of particular research interest focuses on the outer membrane of the bacterium, which contains numerous outer surface proteins (known as Osp) that may hold a clue to the development of a vaccine.

When *Bb* was first discovered, scientists assumed that only one strain existed. But so many strains, differentiated by slight changes in composition, have since been identified that sub-categories, called genospecies, have been created. The overall category is called *Bb sensu lato*. Most *Bb* strains in North America are similar and in research circles are now referred to as *Borrelia burgdorferi sensu stricto.* The strains in Europe and Asia show some marked differences and are designated by several group names. In Europe, the groups *Bb sensu stricto, B. garinii* and *B. afzelii* are found. *B. garinii* and *B. afzelii* have also been identified in Asia. Recently, *B. japonica* and *B. miyamoto* have been named for strains found in Japan. New strains are continually being identified. The outer membrane varies from one bacterial strain to another. There may also be differences in how each strain produces disease and responds to antibiotics. Different strains also require different cultures in which to reproduce and form different colony structures. A new *Borrelia, B. lonestari* sp. nov., has also been identified in the United States.

The Lyme bacterium is intracellular, meaning that it is able to invade human cells. This was first demonstrated in 1987, but it was not until 1995 that a videotape was actually taken through an electron microscope and showed the bacterium invading a cell of the human immune system, dividing, and then breaking down the cell wall, causing the membrane to collapse around itself. That creates a perfect camouflage mechanism that prevents the immune system from recognizing the presence of the foreign invader. This offers one explanation as to why a blood test for antibodies to *Bb* may prove negative even when bacteria are actually present. More importantly, it suggests why certain classes of antibiotics, such as penicillins and cephalosporins, which circulate in body fluids but do not penetrate the cells, may not successfully kill all of the bacteria.

Another fact with implications for treatment involves reproduction time. A strep throat bacterium reproduces every 20 to 30 minutes and is treated through 480 reproductive cycles. It takes the Lyme pathogen 48 hours to first reproduce in culture medium. Once it has adapted to the culture, reproduction time increases to a still relatively slow rate of every 7 hours. Treating Lyme disease through the same number of cycles would require 5 months to 2½ years of antibiotics. Lyme disease almost surely requires treatment for more than just 1 or 2 weeks.

The chromosomal structure of *Borrelia burgdorferi* also intrigues researchers. While lower forms of life (such as bacteria and viruses) have circular chromosomes, *Bb* has linear chromosomes. This differs from all other bacteria and makes *Bb* closer to higher life forms, such as plants and animals, which have a double helix structure. The evolution of this chromosomal structure and its significance remain uncertain.

Another avenue of interest is a substance known as a "bleb." These are usually released in mass, and resemble a string of pearls. Blebs are actually DNA (genetic codes) enclosed in a membrane. They may be the substance that causes the body's inflammatory response to infection. A test to determine whether the presence of Lyme disease could accurately be predicted by measuring blebs has yielded promising results but has yet to be moved to commercial use.

ARE TICK BITES THE ONLY SOURCE OF LYME DISEASE?

There has been only a very limited amount of research into the ways by which the *Bb* spirochete may be transmitted other than through a tick bite. Indeed, this is such a controversial area that most researchers are reluctant even to probe into it. My belief, however, is that what we don't know *can* hurt us, and it makes sense to explore all possible avenues, especially those described here.

Pregnancy and Breastfeeding

The possibility that tick-borne diseases can be transmitted to the developing fetus during pregnancy is one of the most wrenching issues we deal with at the Lyme Disease Foundation. My own personal tragedy, and the loss of my infant son, galvanized me into action, and my concern for other pregnant women continues to spur me on. Because 20% of women in the United States and Europe are of childbearing age, the risks of Lyme disease demand our attention (see Figure 1–5).

Although data on this subject remain scarce, research demonstrates that a pregnant woman who becomes infected with *Bb* and does not receive prompt

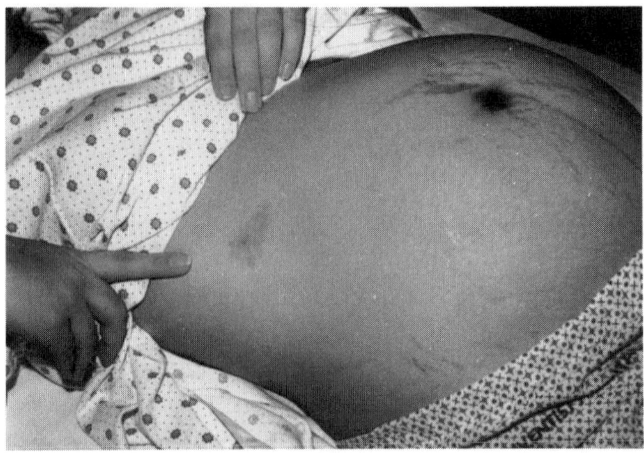

Figure 1–5. Pregnant woman with tick bite. Reprinted by permission of M. Patmas, M.D.

antibiotic treatment can transmit the bacteria through the bloodstream to her fetus, with potentially dire consequences. Rest assured, however: Pregnant women who do receive appropriate treatment generally do very nicely.

In 1985, *The Annals of Internal Medicine* became the first medical journal to publish an article describing a case of maternal/fetal transmission that resulted in the death of an infant. In another study, discussed by T. Gardner in a chapter of *Infectious Diseases of the Fetus and Newborn,* 161 cases of Lyme disease during pregnancy were reviewed, and some sort of adverse outcomes were found in 46 of them. The risk to fetal development was highest if a woman became infected during her first trimester of pregnancy. The author concludes that adverse outcomes (which may be relatively mild or very severe) occur 63% of the time in the first trimester, 38% of the time in the second trimester, and just 10% of the time if infection occurs in the final trimester.

Documented adverse outcomes include a risk of retarded growth, respiratory distress, eye problems, brain infection, heart abnormalities, and damage to other organs. On rare occasions, fetal malformations, miscarriage, and early infant death have been linked to Lyme disease, and the spirochete has been recovered from both fetal remains and placental tissue. In addition, the possibility that Lyme disease in pregnancy is linked to subsequent developmental delays and learning disabilities has been documented.

Also of concern are the results of recent studies suggesting that it may be possible to transmit infection in breast milk. A study published by B. Schmidt in 1995 detected *Bb* in two samples of milk provided by lactating women with Lyme disease. A year later, a study conducted in mice also proved that early breast milk from an infected mother could transmit the infection to the newborn. This experiment raises questions about earlier assumptions that an infant's stomach acid would most likely kill any Lyme bacteria.

While every pregnant and nursing woman needs to take special precautions to prevent a tick bite, there is no surefire way to stay safe. Securing an accurate diagnosis can be challenging, especially given the many discomforts that are part of almost any pregnancy. If you are pregnant and are bitten by a tick, prompt antibiotic treatment is crucial, even if you don't know for sure that you have been infected.

Human and Animal Contact

Scientists have long assumed that transmission through sexual contact does not occur and that if two members of a couple were infected, it was because each had been separately bitten by a tick. Until 1996, however, no one had actually bothered to study this important issue. That year, a researcher at the State University of New York at Buffalo challenged prevailing wisdom with a study demonstrating that infected male mice can transmit *Bb* to the females and their offspring. It is an understatement to say that this area needs more investigation.

Lyme disease can also be transmitted through animal urine, although any bacteria generally die as soon as the urine dries. Nonetheless, at least two scientists I know acquired infection directly from infected animals. In one instance, the scientist was splashed in the eye with an infected animal's urine. The other scientist was bitten on the hand by an infected animal, which then urinated on the open wound. Uninfected animals can also contract Lyme disease by ingesting the urine of infected animals. To protect themselves, veterinarians wear arm-length gloves and protective glasses when collecting large-animal urine samples.

There is absolutely no evidence that casual contact, such as touching, kissing, or changing diapers, transmits the Lyme bacterium. Nor does it seem likely that you will get Lyme disease from handling an infected dog or cat. My family had five pets with Lyme disease, but my husband, who tended them all, never became infected.

Other Possible Modes of Transmission

Insects

Although an insect can become infected with the Lyme bacterium, the extent to which it is able to retain it or to transmit it to another host is unclear. Deerflies, horseflies, and fleas, all of which feed on blood and bite human beings repetitively, have been implicated in the cycle that leads to disease, although they appear to be minor sources of infection. Of particular concern is the ubiquitous mosquito, which can definitely become infected with *Bb*. Because the bacteria live only a few hours in a mosquito's body, the likelihood of transmission appears remote, but no one has actually designed a transmission study with mosquitoes. It urgently needs to be done.

Blood Products

Because the Lyme bacteria travel through the bloodstream, researchers have been concerned about the possibility of contamination in the nation's blood supply. In 1987, Drs. Richard Fister and Richard Tilton, researchers at the

University of Connecticut School of Medicine, working in collaboration with the American Red Cross, spiked donated blood with *Bb* and then processed and froze the sample, as would be done prior to use in a transfusion. When the blood was reprocessed as if for use, the researchers were able to recover live, active spirochetes. When a CDC researcher showed similar results in 1990, the American Red Cross finally began to exclude patients with active Lyme disease as donors.

The issue of whether menstrual blood could contain Lyme bacteria has not been studied at all.

Transplants

According to *Aspects of Lyme Borreliosis,* a book published in Heidelberg, Germany, by Springer-Verlag in 1993, a small body of research suggests that infection can be transplanted. In 1954, a German dermatologist, Hans Götz, was able to transplant infection into healthy volunteers from a biopsy of a Lyme disease skin condition known as ACA (more on this later). Similar experiments with comparable findings were conducted in 1955 by E. Binder and in 1958 by J. Paschoud regarding other Lyme disease skin conditions. Since then, no investigations have been done in this important area, although the Lyme bacterium has been cultured from a vast array of transplantable human tissue. There is little reason to think transmission cannot occur through a transplant.

OTHER TICK-BORNE DISORDERS

Lyme disease is not the only disorder carried by infected ticks, only the best-known and most common one. At least eight other serious illnesses in humans in the United States have been linked to a tick bite. The pathogens for at least two of them—babesiosis and ehrlichiosis—are carried by some of the same ticks that carry *Bb,* therefore more than one infection can be transmitted through a single bite. For example, one study has shown that half the ticks in Connecticut that transmit Lyme disease are also infected with the pathogen that causes human granulocytic ehrlichiosis. (See chapter 7 for more information about these eight diseases.)

1. Babesiosis (pronounced bab-ee-see-O-sis), caused by various species of *Babesia* (protozoans), invades, infects, and kills the red blood cells. It begins with flu-like symptoms but can ultimately affect the liver, kidneys, and

spleen. The illness appears to suppress the immune system and can be fatal in elderly people. Antibiotic and quinine treatments are generally effective, although a blood transfusion may be required.

2. Ehrlichiosis (pronounced err-lick-ee-O-sis), caused by a species of parasite called *ehrlichia* (a type of *rickettsia*), infects the white blood cells and in some instances becomes severely debilitating within hours of a tick bite. Two classes of disease—human monocytic ehrlichiosis (HME) and human granulocytic ehrlichiosis (HGE)—were officially discovered only a few years ago, but scientists now realize that they have been around for years. While antibiotics are generally effective if administered promptly, death may result from infection, generally when it is not recognized early or is treated inappropriately.

3. Rocky Mountain spotted fever (RMSF), also called tick-borne typhus, is caused by the *Rickettsia rickettsii* parasite. A classic trio of symptoms—fever, headache, especially behind the eyes, and a hallmark skin rash that resembles the shape of measles—should alert your physician to this infection. Because serious complications, including gangrene, kidney failure, and central nervous system problems can occur, the disease can be fatal if it is not treated early and adequately. A promising vaccine was formerly in development, but research was abandoned because the demand was not deemed sufficient.

4. Colorado tick fever, caused by a type of virus known as a reovirus, begins with a sudden onset of very high fever, followed by a period of remission, and then another bout of fever. Heart problems and other complications can occur. No antiviral therapy is available, but individual symptoms may respond to supportive treatment.

5. Tularemia has been called a plague-like disease. It is caused by the *Francisella tularensis* bacterium. Painful and swollen lymph nodes are the most characteristic symptoms. They can break through the skin and develop into abscesses. Antibiotics are generally effective, although relapses can occur. A vaccine has been developed, but it is not 100% protective, and its use is restricted to people in high-risk jobs.

6. Relapsing fever, caused by at least three different *Borrelia* spirochetes, is characterized by high fever, followed by sudden chills. It can involve multiple organ problems. Eye inflammation, jaundice, and coughing commonly occur. The disease does respond to antibiotics. Because relapsing fever and Lyme disease are caused by related bacteria, scientists often turn to the published literature about relapsing fever for clues about Lyme.

7. Powassan encephalitis is caused by the flavivirus. It results in neu-

rologic illness, including seizures, paralysis, and brain inflammation. This is a rare disease, one for which there is no effective treatment.

8. Tick paralysis is a potentially fatal reaction to a toxin secreted in the saliva of female ticks late in their feeding. Paralysis generally begins in the legs and spreads throughout the body within hours. If the tick is found and removed, recovery is swift; otherwise, nerve damage and respiratory failure can occur.

An important fringe benefit of recent Lyme disease research has been to put a spotlight on all tick-borne disorders. Hopefully, new knowledge about the life cycle of ticks and how they spread infection will benefit not only people with Lyme disease but also anyone at risk for any illness carried by tick parasites.

2

UNDERSTANDING TICKS AND THE INFECTIONS THEY SPREAD

What are ticks? It depends on who you ask. My admittedly biased opinion is that they are rather creepy bugs that serve no useful purpose. I am in good company with my view. Even the Greek philosopher Aristotle agreed, writing in the fourth century B.C. that ticks are "disgusting parasitic animals."

But entomologists, the researchers who make a life's work out of studying these critters, have a more sophisticated and scientific description. They have classified ticks as arthropods, a biological grouping of invertebrate animals with jointed legs, segmented bodies, and an external skeleton. Within the arthropod classification, ticks are arachnids, a subcategory that also includes spiders and mites but not insects. There are three main differences between insects and ticks:

- Insects have three body segments, but ticks have two.
- Adult insects have six legs, but adult ticks have eight.
- Insects have wings or antennae, but ticks do not.

More than 850 species of ticks have been identified worldwide. Of these, 100 are capable of transmitting pathogens, such as bacteria, viruses, and protozoa, as well as toxins, to human beings. In North America, 5 species of ticks—the *Amblyomma, Dermacentor, Ixodes, Ornithodoros,* and *Rhipicephalus*—can transmit these illness-causing agents to humans.

"Ticks are cesspools," comments Ben Luft, M.D., chairman of the Department of Medicine at the School of Medicine on the Stony Brook campus of the State University of New York, in a colorful description of how much undesirable baggage ticks carry with them.

ARE TICKS ON THE RISE?

Since I became a Lyme disease educator, I cannot count the number of times I have been asked, "Where have all of these ticks come from? I don't remember such problems when I was growing up."

Actually, ticks aren't new at all, nor are the diseases they spread. Many of the diseases making headlines now are newly named, but their trail of destruction has ancient origins. Indeed, ticks may have been a threat to the public health as long as we have coexisted on this planet. James Oliver, Ph.D., of the Institute of Arthropodology and Parasitology at Georgia Southern University, which is home to the National Tick Collection of the Smithsonian Museum, theorizes that ticks have existed on Earth for more than 400 million years, originally feeding on amphibians. Although the historical record is not definitive, the biblical plague visited on the cattle of Pharaoh Rameses II may actually have been the tick-borne protozoan infection now called Texas cattle fever. More certain is that in 200 B.C., M. Procius Cato, a Roman statesman and philosopher, spoke of an ideal time in the future when "there will be no sores and the wool will be plentiful and in better condition and the ticks will not be troublesome."

In part, the perception that tick-borne diseases are on the rise may reflect increasing scientific interest in the subject. Since the Lyme Disease Foundation began its advocacy work in 1988, substantial public and private investments have been made in research, and the media has finally begun to pay attention to the issue.

But the chances of being bitten by an infected tick may also have increased in real terms. As we move into the twenty-first century, more people are traveling to new areas, exploring remote locations, building on previously undisturbed land, and enjoying vigorous outdoor activities—all of which increase exposure to tick-spread diseases. A number of other forces have converged to increase the threats of ticks and tick-borne disorders:

- **Changing Landscape:** Over the past 30 or 40 years, the American landscape has changed markedly. A look back at history is instructive. In his book *Ticks: And What You Can Do about Them,* Roger Drummond, Ph.D., quotes from an eighteenth-century traveler in New York State who says that it is impossible to sit outdoors without being attacked by an army of ticks. By the late nineteenth century, the landscape had been so dramatically altered—with wilderness giving way to farms and grazing lands—that another visitor to the same area wrote that the common tick had nearly become extinct. Now, with much of the U.S. Northeast more heavily forested again, ticks have made a remarkable comeback.

Shifts in the pattern of Rocky Mountain spotted fever illustrate the results of ecological disruptions. In the 1940s, the disease was primarily found in the Rocky Mountains, but as settlers cleared the land for pastures, ticks were pushed out, and the number of cases dropped dramatically. Meanwhile, east of the Rockies, farmland was increasingly giving way to forests, and suburban terrain was being planted with tick-attracting shrubbery, with a resulting rise in the incidence of disease. By 1964, more than 90% of Rocky Mountain spotted fever cases were being reported in the eastern United States.

- **Regulatory and Aesthetic Shifts:** Environmental regulations have also encouraged the survival and spread of tick populations. Restrictions on leaf burning and the decreasing use of pesticides, especially DDT, have allowed more ticks to survive. Likewise, short hunting seasons mean longer lives for many of the animals that are host to ticks.

Aesthetic preferences and a greater interest in the natural environment can also create a health threat. Some homeowners have removed property fences to encourage grazing wildlife. Unfortunately, that is an open invitation to the ticks that travel with the wildlife. While the animals are making a meal out of your plantings, or even your dinner scraps, the ticks may be making a meal out of you! Likewise, the lush forestland and undergrowth that surround homes in many communities create an ideal environment for tick hosts.

- **Patterns of Antibiotic Use:** According to the medical textbook *Principles and Practice of Infectious Diseases,* the incidence of Rocky Mountain spotted fever declined when antibiotics, especially tetracycline, came into widespread use. In the 1970s, when the popularity of tetracycline declined, the number of Rocky Mountain spotted fever cases began to rise. The implication is that prior to the 1970s, a number of undiagnosed cases of the disease were being effectively treated when tetracycline was prescribed for other purposes.

THE LIFE CYCLE OF A TICK

Ticks have four life stages: the egg, larva, nymph, and adult. The larva, nymph, and adult are the active stages when a tick may be able to spread infection.

A tick moves from one active stage to another by molting, meaning that it sheds its old skeleton and develops a new one to match its new body structure. Molting occurs immediately after feeding and usually takes a few days. The entire life cycle of a tick averages 2 years, but it may be as short as 1 year or as long as 3, depending on the availability of food.

1. **Egg:** The life cycle begins as the adult female tick mates with a male. After mating, the female requires a blood meal in order to produce eggs, which she may lay either in a single group or in several batches.
2. **Larva:** The larva that hatches from the egg, sometimes referred to as a "seed" tick, has six legs and no gender differentiation—males cannot be distinguished from females. The larva tick is no larger than the period at the end of this sentence but triples in size when fed. After consuming a blood meal, the larva molts into a nymph. An unfed larva has been known to survive as long as 1 year.
3. **Nymph:** The nymph has eight legs and is still not differentiated by gender. The body and legs of a nymph tick would fill a small letter "o" on this page. Again, the tick consumes a blood meal and then molts into an adult.
4. **Adult:** Adult males and females have eight legs and are distinctive in gender. The adult tick's body would cover the capital letter "O" and enlarges to about the size of a plump raisin after feeding. An adult female usually lays eggs and dies soon afterwards, although unfed adult soft ticks can survive as long as 18 years.

One of the great ironies about ticks is that even though they have very primitive immune systems, only a very few species become infected with the *Borrelia burgdorferi* spirochete. Humans, by contrast, have tremendously advanced immune systems yet are highly vulnerable to infection.

HOW TICKS FEED

As parasites, ticks are totally dependent for food on the blood and tissue fluid of a host, which may be a person, a wild or domestic animal, a bird, or a reptile. Woods and forests are the preferred tick habitat, but they also live in ground vegetation, grassy areas, meadows, weeds, leaf litter, caves, and sometimes cabins that are left empty for the winter. Ticks thrive in humidity and tend to live in close proximity to potential hosts. This frequently places them in the dense, overgrown area between a manicured lawn and a forest, an area called the "ecotone," which is a transition between neighboring ecological communities.

Ticks are positively geotrophic, meaning that they instinctively move in opposition to gravity, typically seek food by climbing upwards on brush, and, in behavior that is called "questing," use one set of legs to grab onto a host as it brushes past. A tick responds to the stimuli of exhaled carbon dioxide, scent, and body heat, among other factors, to find a host. A tick rarely moves higher than three feet above ground level and cannot jump or fly.

Most ticks prefer to feed on nonhuman mammals, especially deer, mice, chipmunks, and rabbits, and on birds, which can maintain infection for life. A study conducted by the Connecticut Agricultural Experiment Station demonstrated that more than thirty-five species of birds, including blue jays, robins, wrens, finches, and sparrows, carry ticks infected with *Bb* in Connecticut alone. And the Gundersen Medical Foundation in Wisconsin showed that thirty species of birds carried an average of 3.5 ticks per bird.

Regardless of preference, a tick will feed where it can. If you are passing by at the right time, you may well be considered an adequate host. The mouthparts consist of a hypostome and the chelicerae. The hypostome is a hollow tube with harpoonlike reverse barbs on the outside that are designed to secure attachment in the skin. The chelicerae are retractable cutting edges that are located inside the tube of the hypostome. The chelicerae are designed to slice the tick's way into the host's skin. In the course of attaching to your body, the tick releases an anesthetic to prevent you from feeling its bite. As it begins to feed, a cementlike substance is secreted from the saliva to enhance its capacity to adhere to your skin. Chemicals that act as anticoagulants and antiinflammatories are also released to keep your blood flowing and to provide the tick better access to its food. This is the point at which a tick may transmit a pathogen or acquire one from an infected host. While the male adult tick may take a blood meal, the female adult is most likely to transmit disease because it must feed before it can lay eggs, and it is capable of feeding for a longer period of time.

Only in 1893, after 20 years of investigation, did scientists actually prove that ticks could transmit infectious agents such as bacteria, viruses, or protozoa from one animal to another. The first confirmed tick-borne disease was called Texas cattle fever. Today, we know that ticks transmit more kinds of microorganisms than any other arthropod, including mosquitoes, although 90% of all ticks transmit disease only to a specific nonhuman, nonpet wild animal host. The main public health concern centers on the remaining 10%.

If a tick feeds on an infected host, the tick itself may become infected and then may pass the infection along to the next host. However, not every microorganism acquired by a tick during feeding can be deposited into fluids of another host. For example, dog ticks can become infected with the Lyme disease bacterium, but they cannot maintain the infection as they molt and therefore cannot transmit it to others. A tick that can maintain infection as it molts from one stage to another is called a "competent vector."

From a pathogen's perspective, rodents are the ideal hosts, because they maintain infectious agents in their bloodstream throughout their lives, which are long enough to infect several generations of new ticks. Humans, by contrast, are merely incidental hosts for the pathogens. In fact, for tick-spread infectious agents they are a "dead end host" because an uninfected tick usually does not acquire and pass along an infectious agent from an infected human being.

WILL I BECOME ILL FROM A TICK BITE?

While it is impossible to say whether or not the tick that bites you is infected, and with what pathogen, a tick bite is always of concern. Many factors increase or lessen the likelihood that you will become ill and may also determine how sick you will become and how quickly the infection takes to manifest itself.

Local Infection Rates

You are more likely to get infected if there is a high percentage of infected ticks in your community, but accurate statistics are very hard to come by. Infection rates change from year to year, from one region of the country to another, and sometimes even within the same county, so estimates depend in part on where and when researchers have gone to collect their ticks. Because some communities are much more rigorous than others about studying rates of tick infection, anecdotal reports from neighbors and the local media may actually give you the best sense of the problem. If someone tries to assure you that there is no disease in your community, ask careful questions. Chances are that no one has actually made a thorough study of the issue. More surprising is the fact that, according to Dr. Willy Burgdorfer, *all* ticks carry the bacterium known as *Wolbachia,* a *rickettsia*. This pathogen's role in human or animal disease is unknown.

Number of Tick Bites

The more often you are bitten by a tick, the greater your chance of eventually becoming infected. Say, for example, that the local tick infection rate is 20%. If you receive five tick bites, odds are good that one of them is going to make you ill. Some physicians are a bit dismissive when their patients report a bite, saying, "It's only a single tick bite. You are probably fine." Unfortunately, there is no way to be sure.

Strain Differences

Strain variations exist even among pathogens that cause the same disease. Some are more virulent than others and therefore more likely to produce infection. There is also some evidence that different strains of disease-causing agents result in different symptoms. Differing effects of the 300 strains of *Bb* are an area of research interest, especially the effects of coinfections with multiple strains.

Inoculating Dose

In order for infection to establish itself, a minimum number of pathogens must be introduced into the host, as Lyme disease researchers have discovered by injecting various levels of *Bb* into animals. We don't yet know exactly what dosage results in human infection. Research is complicated by the fact that certain ticks may be able to transmit disease more readily than others. The level of pathogens in your body may also determine how sick you become. In some instances, a patient may develop a level of infection subtle enough not to notice yet destructive enough to cause slow but progressive deterioration.

Feeding Time

In general, the longer an infective tick is attached to a host and feeding, the greater the chance that the host will become infected. But again, there are significant differences from one type of tick to another. Soft *Ornithodoros* ticks can transmit pathogens in as little as 15 minutes, while other tick species need to be attached for 1 or 2 days to cause infection. One fact holds in every case—the sooner you *properly* remove a tick, the less likely you are to become infected.

Host Immunity

In a host with a compromised immune system, disease generally occurs more rapidly and can be much more severe. This is especially evident with human granulocytic ehrlichiosis, which is much more likely to kill someone with an immunocompromised system than someone who has been healthy.

Systemic versus Local Infection

One of the least studied and most intriguing features of ticks is that some are systemically infected with pathogens found both in the midgut and salivary glands while others are infected locally, with the agent present only in the midgut. The distinction is very important. In Lyme disease, for example, a systemically infected tick may be able to transmit a pathogen in only a few hours whereas a tick with a localized infection may take 24 to 48 hours to move the bacteria into its salivary glands so that transmission can occur.

TYPES OF TICKS

Ticks are divided into two broad categories—hard ticks, which have either a partial or full shell over their backs, and soft ticks, which have no hard shell at all—and differ in their longevity, reproductive powers, preferred habitat, response to environmental conditions, and ability to transmit pathogens to animals. Most ticks look very similar to the untrained eye but have a number of differing characteristics that make them biologically distinct.

Hard Ticks

There are about 650 different types of hard ticks in the world, all of which share the following characteristics:

- Larvae feed once and molt into nymphs; nymphs feed once and molt into adults; adults feed once.

- Once a tick pierces the skin and starts feeding, it continues to feed to completion—which can be from several days to a week.
- The tick's mouthparts extend beyond the body, making them visible when viewed from above.
- Larvae and nymphs generally feed on smaller animals. Adults generally feed on larger animals, although ticks in any stage can bite human beings.
- They have a smooth-looking skeleton, also called a shield.
- The female has a partial shield covering its back, and the male has a full shield over its entire back.
- The skeletal shield of the adult male limits its capacity to expand during feeding.
- The female adult expands like an accordion as it becomes full, sometimes reaching as much as ten times its original size.
- The female lays anywhere from 200 to 23,000 eggs, depending on the species, usually in one batch, and then dies.
- They live outdoors.

Soft Ticks

There are approximately 150 species of soft ticks in the world, but only members of the *Ornithodoros* species, which spread pathogens causing relapsing fever, are of concern to humans in North America. All soft ticks share the following characteristics:

- They feed multiple times during each life stage.
- They molt between three and seven times at each stage.
- They feed for brief periods of time, generally between 12 minutes and 1 hour.
- The tick's mouthparts are not visible from above.
- They do not have a hard shell.
- Male and female look alike.
- They look leathery and wrinkled.
- The bodies of both male and female soft expand equally during feeding.
- The adult female may lay a total of 50 to 200 eggs in its lifetime.
- They live in sheltered places, including caves, cabins, and the nests or burrows of other animals.

Eight Ticks to Blame

These are the eight ticks that have been reported to carry disease-causing pathogens in North America:

1. **Black-legged tick** (*Ixodes scapularis*)
   formerly called the deer or bear tick

   Disease transmission: The black-legged tick transmits the pathogens for Lyme disease, babesiosis, human granulocytic ehrlichiosis, and Powassan encephalitis. It is also believed, though not proved, to transmit pathogens for tularemia. Some researchers believe it causes tick paralysis.

   Geographic range: This tick can be found across eastern North America, extending as far as the Midwest.

   Life cycle: Larvae, which are black and slightly transparent, are active in the summer and fall, and overwinter in the soil. Because of their transparency, they may look flesh-toned when feeding. Nymphs are black and are active all summer. A nymph that is attached and feeding looks somewhat like a freckle with legs. Adults feed in the fall, on warm winter days, and into the early spring. During the cooler parts of winter, they lodge in the soil and leaf litter.

   Gender markings: The adult female has a solid black shield on a portion of its back and a red-brick-colored body, which changes to taupe after feeding. The adult male coloring ranges from dark brick to black, and a full black shield covers its entire back.

   Other facts: These ticks are visually identical to the Western black-legged tick. The adult female lays between 250 and 1,000 eggs at one time.

2. **Western black-legged tick** (*Ixodes pacificus*)

   Disease transmission: The Western black-legged tick (Figure 2–1) transmits the pathogens for Lyme disease and probably human monocytic ehrlichiosis (HME), human granulocytic ehrlichiosis (HGE), and possibly tularemia. It has also been associated with a new, unnamed California strain of *Babesia* and is believed to cause tick paralysis.

   Geographic range: These ticks can be found along the Pacific Coast

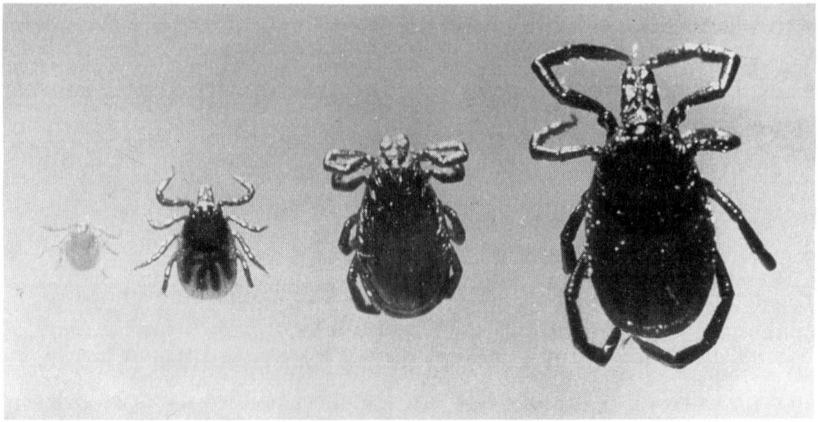

Figure 2–1. Western black-legged ticks: (left to right) larva, nymph, male, and female. Reprinted by permission of R. Lane, Ph.D.

of the United States and Canada and as far inland as Nevada, Utah, and Idaho.

Life cycle: Like the black-legged tick, larvae are black and because of their transparency may appear flesh-colored when attached and feeding. They are active in the late spring and summer. Nymphs are black, and they, too, are active all summer. Adult ticks are active from November through May.

Gender markings: The adult female has a solid black shield covering a portion of its back and is a reddish brick in color, becoming taupe after feeding. The adult male coloring ranges from dark red brick to black, and a full shield covers its back.

Other facts: These ticks look virtually identical to the black-legged tick, and like them, the adult female lays between 250 and 1,000 eggs at one time.

3. **Lone star tick** (*Amblyomma americanum*)

   Disease transmission: The lone star tick (Figure 2–2) transmits Lyme disease-causing *Bb* and is a major vector of tularemia. Recently, it was found to carry another *Borrelia* species associated with Lyme-like symptoms. The lone star tick is also a major vector for the HME pathogen, and in the Southeast it is suspected of transmitting the *Ehrlichia* species that causes HGE. Lone star ticks have also been found to be infected with *Rickettsia rickettsii,* which causes Rocky

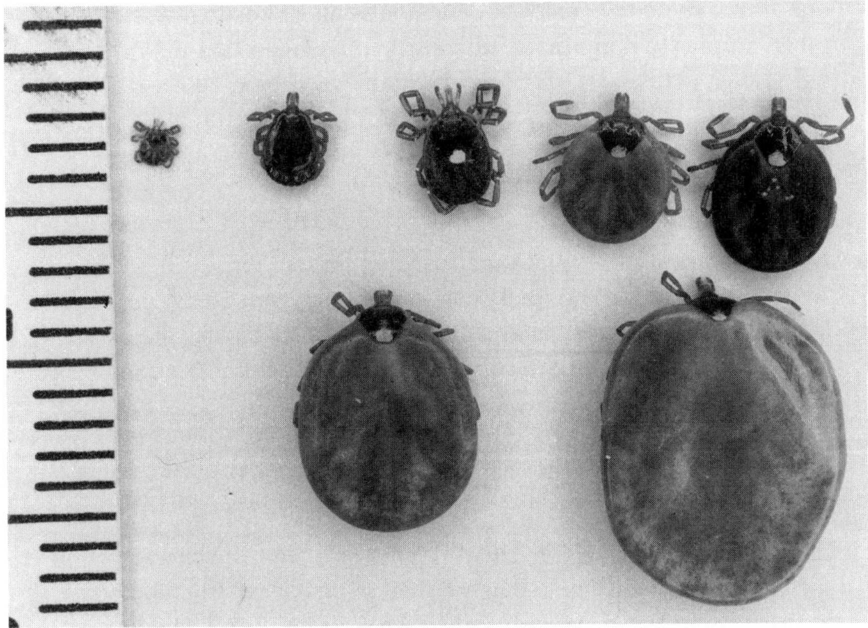

Figure 2–2. Lone star ticks by a cm measure: (left to right, top) nymph, male, unengorged female, and two females in different stages of engorgement; (left to right, bottom) two females in further stages of engorgement. Reprinted by permission of D. Feir, Ph.D.

Mountain spotted fever, but it is not certain whether it is capable of transmitting infection to a human host. The bite of the tick larva can also cause itching or a rash. The female may cause tick paralysis as well.

Geographic range: This tick can be found within the rectangle encompassing Rhode Island, Florida, Texas, and Iowa.

Life cycle: The larvae hatch in late summer and range from yellow to reddish-brown in color. Nymphs appear from spring to fall and winter over in the soil. Adults are active in the spring and summer, also wintering over buried in the soil.

Gender markings: The female adult has a reddish-brown body and a unique white dot on the black shield that covers a portion of its back. The male adult has a reddish-brown body and a lacy, pale white marking on the outer fringe of its back.

Other facts: The lone star tick is aggressive and seeks out human beings to bite. The adult female lays from 15,000 to 23,000 eggs at one time. The adult male lone star tick can be distinguished from other, similar-looking ticks, such as the American dog tick, the Pacific Coast tick, and the Rocky Mountain wood tick, because it does not have the same chalklike markings down the top of its body.

4. **American dog tick** (*Dermacentor variabilis*)
also called a wood tick

Disease transmission: The American dog tick (Figure 2–3), which is sometimes called a wood tick, can transmit the pathogens for Rocky Mountain spotted fever, human monocytic ehrlichiosis (major vector), and tularemia. It is also a suspected vector for HGE-causing organisms and can cause tick paralysis.

Geographic range: This tick is found in all 50 states and all of the Canadian provinces, although it appears to be absent from the Rocky Mountain area.

Life cycle: Larvae, which are yellowish in color, are active in the spring and summer. The reddish-brown nymphs are generally active in the summer but may survive the winter in the soil. The adults are

Figure 2–3. Left group shows American dog ticks: (left to right) engorged adult female, adult female, adult male. Right group shows black-legged ticks: (top row, left to right) three larvae, engorged larva, two nymphs; (bottom row, left to right) engorged adult female, adult female, adult male. Reprinted by permission of Karen Vanderhoof-Forschner.

active from spring through fall and may winter over under leaf litter or in the ground, becoming active again in warm weather.

Gender markings: The adult female American dog tick has two wavy, silvery, chalklike marks resembling an "s" that run lengthwise down the shield that covers a portion of its back. The rest of the body is reddish-brown. The adult male has similar markings down the full length of the top of its body and is a somewhat darker reddish-brown than the female.

Other facts: This tick earned its name because it prefers to feed on dogs, but it will also feed on other animals and humans. It looks very similar to the Pacific Coast and Rocky Mountain wood ticks, both members of the same species. (See Figure 2–4.) The adult female lays between 3,000 and 6,000 eggs at one time.

5. **Pacific Coast tick** (*Dermacentor occidentalis*)

Disease transmission: The Pacific Coast tick transmits the pathogens for Rocky Mountain spotted fever, tularemia, and probably Colorado tick fever.

Geographic range: This tick is found in California, Oregon, and as far east as Nevada.

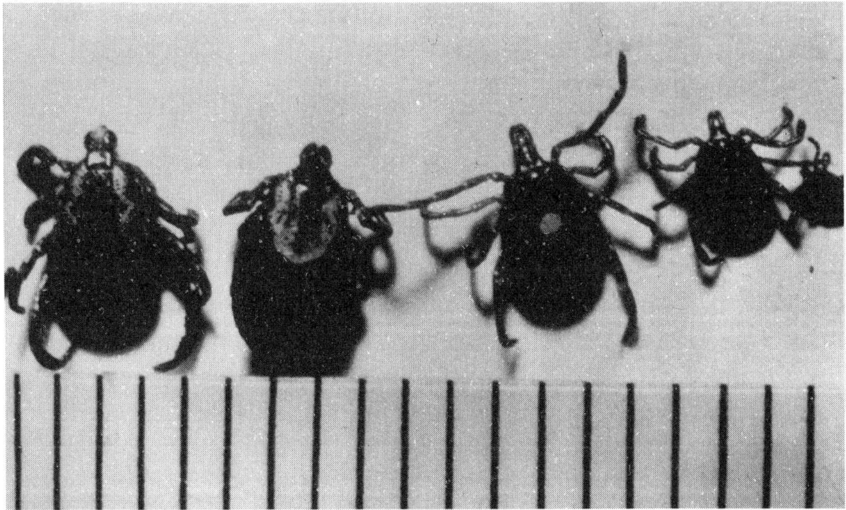

Figure 2–4. Adult male and adult female American dog ticks are shown on the left; adult female, adult male, and nymph lone star ticks are shown on the right. Reprinted by permission of D. Feir, Ph.D.

Life cycle: Larva and nymph ticks are dark brown and are active in the late spring and early summer. Adults are active in the late spring.

Gender markings: The female adult Pacific Coast tick has two wavy-shaped silvery chalk-like marks resembling an "s" that run lengthwise down the shield that covers a portion of its back. The rest of the body is reddish-brown. The male adult has similar markings down its entire back and is dark reddish-brown.

Other facts: The Pacific Coast ticks look very similar to the American dog and Rocky Mountain wood ticks, both members of the same species. The adult female lays between 3,000 and 6,000 eggs at one time.

6. **Rocky Mountain wood tick** (*Dermacentor andersoni*)

Disease transmission: The Rocky Mountain wood tick is the primary cause of tick paralysis in Canada and the United States and also transmits the pathogens for Rocky Mountain spotted fever, tularemia, Colorado tick fever, and Powassan encephalitis.

Geographic range: This tick is found from the Rocky Mountain states to the Pacific Coast and into Alaska, as well as across the central states to North Dakota and Nebraska.

Life cycle: The larvae are reddish-brown and are active in the summer. The nymphs are dark reddish-brown and are active in the spring and summer, wintering over buried in the soil. In the adult stage, the ticks are active primarily in the early summer.

Gender markings: The adult female Rocky Mountain wood tick has two wavy-shaped silvery chalk-like marks resembling an "s" that run lengthwise down the shield that covers a portions of its back. The rest of the body is reddish-brown. The adult male has similar markings down the full length of the top of its body and is dark reddish-brown.

Other facts: The Rocky Mountain wood tick looks very similar to the American dog and Pacific Coast ticks, both members of the same species, and can also be confused with the male lone star tick, except the latter does not have chalk-like marks. The adult female lays between 3,000 and 6,000 eggs at one time.

7. **Brown dog tick** (*Rhipicephalus sanguineus*)

Disease transmission: The brown dog tick transmits the bacterium

Ehrlichia canis to dogs and can infect human beings with *Ehrlichia chaffeensis,* which causes HME.

Geographic range: This tick is found in every state, including Alaska and Hawaii, and often lives indoors. The brown dog tick usually bites dogs in Canada and most of the United States, but it has been known to bite people in the very southern parts of the United States, including Texas, as well as in Mexico and Europe.

Life cycle: The larva and nymph ticks are dark brown. The adult male is also uniformly dark brown, while the adult female is brown with a partial shield that is darker than the rest of its body. After feeding, the adult female turns a dark gray.

Other facts: The female brown dog tick lays between 3,000 and 6,000 eggs at one time.

8. **Relapsing fever ticks** (*Ornithodoros hermsi, Ornithodoros turicata, Ornithodoros talaje*)

   These are the only soft ticks in North America able to infect humans. All three of these tick species look similar, but each one carries a different and specific strain of relapsing fever bacterium. They are found in southwestern and south-central states. They live for 5 to 10 years without feeding. During the larva and nymph stages, these ticks are gray; as adults, they are grayish to pale blue in color and have a wrinkled body.

3

THE HISTORY OF LYME DISEASE

A popular misperception—born at least in part of academic rivalries, the pursuit of research grants, and the quest for fame—has it that Lyme disease was discovered in 1977 in Lyme, Connecticut. That story has been repeated so many times in the press that most Americans probably accept it as gospel—but it is just not true.

The medical literature is actually rich with more than a century of writing about the condition, although much of it has been published only in Europe. By 1945, researchers had already described much about what we now call Lyme disease—including the variety of skin problems, the common arthritic conditions, and many of the neurological, cardiac, and ophthalmologic problems that suggest the involvement of multiple body systems. Over the years, the complex of signs and symptoms has been referred to in countless ways, from Bannwarth's syndrome, named for a German neurologist, to neuroborreliosis, chronic lymphocytic meningitis, and polymeningoradiculitis. The characteristic skin rash initially carried the names of scientists who had identified it, including erythema chronicum Lipschütz and erythema migrans Afzelius.

By midcentury, some researchers were also convinced that the *Ixodes* tick was involved in transmitting an infectious agent and a handful of scientists considered a spirochete to be the most likely culprit. By then, the disease was widely recognized as serious and known to have multiple stages, beginning with the tick bite and disseminating throughout the body. Antibiotics were being used in treatment, but it had already become apparent that they didn't always effect a cure. Subsequent research has proven that many of the findings of these early researchers were right on target.

Many of the strategies used to safeguard against infection were also described long ago. In the 1970s, Tennessee Valley Authority (TVA) researchers, assisted by the U.S. Department of Agriculture (USDA), began to develop the Integrated Tick Management Program, which combines personal protection, property management, host management, and pesticide use. That strategy is still the preferred way to minimize the dangers of ticks and will be described in detail later in this book.

Because so few of the older research findings have ever been distributed in the United States, I have made a special effort to present here the significant events in the history of Lyme disease. Take a look at the "Timeline" in the appendix for many more details.

PHASE ONE: A CENTURY OF OBSCURITY

The first record of a condition associated with Lyme disease dates back to 1883 in Breslau, Germany, where a physician named Alfred Buchwald described a degenerative skin disorder now known as acrodermatitis chronica atrophicans (ACA). Had German researchers been more media savvy, we might well be talking now about "Breslau disease" or even "Buchwald disease." Instead, we are saddled with a name that has no historical validity and does not offer any hint of disease severity.

Also of interest is the 1909 meeting of the Swedish Society of Dermatology, where a physician named Arvid Afzelius presented research about an expanding, ringlike lesion he had observed (hence the name erythema migrans Afzelius). Afzelius published his work 12 years later and speculated that the rash came from the bite of an *Ixodes* tick. Later, the rash was named erythema chronicum migrans, simplified in the last decade to erythema migrans (EM). It is now recognized as one of the most characteristic signs of Lyme disease, so much so that the rash itself is a diagnostic for Lyme disease.

Associations among many of the symptoms and signs that constitute Lyme disease were already being made in the early part of the twentieth century. The first report of serious joint involvement in patients with ACA came in 1921. A year later, French physicians Charles Garin and Charles Bujadoux observed the link between the EM rash and neurologic problems. Psychiatric symptoms were observed in patients with the EM rash as early as 1930, and four years later, a Swedish researcher reported that some patients with benign lymphocytomas, another skin disorder now linked to Lyme, also had either EM or ACA. Also in 1934, a German scientist described the heart involvement that appeared in patients with both the EM rash and arthritic symptoms.

By midcentury, physicians were experimenting with still-novel antibiotics, including penicillin and tetracycline, and reporting positive results. In 1951, the first report was presented suggesting that after they start treatment, patients sometimes get worse before they get better. Today, we know this as the Jarisch-Herxheimer reaction.

Unfortunately, there were also ignoble moments in early history, with some scientific ideas never receiving a fair hearing. For example, an Austrian dermatologist, B. Lipschütz, wrote at the beginning of the twentieth century: "Perhaps we are dealing with a skin infection caused through the bite of a tick. Therefore, attention should be directed towards microscopic/bacteriologic investigations of the intestinal tract and of the salivary gland secretions of the tick." Alas, "no one followed his suggestion," noted Willy Burgdorfer, discoverer of *Borrelia burgdorferi* (*Bb*), the spirochete named for him.

Another researcher of great repute who was never given his full due was Sven Hellerström of the Karolinska Institute in Sweden. In his work, conducted in close cooperation with Carl Lennhoff, a refugee from Nazi Germany, and Einar Hollstrom, a departmental assistant, Hellerström suggested the spirochetal origin and neurological sequestering of Lyme disease in the 1940s. Unfortunately, "the contributions of Dr. Hellerström and the Europeans in general were largely overlooked in the American literature," wrote Rudolph Scrimenti, associate clinical professor in the Department of Dermatology at the Medical College of Wisconsin, and an expert on Lyme disease skin conditions, in the January 1993 issue of the *Wisconsin Medical Journal.*

If full credit was not always given to the European pioneers, American researchers nonetheless benefitted enormously from their knowledge, which was accruing long before Lyme arthritis was officially discovered. Researchers today still reap the fruits of decisions made long ago, such as preserving ticks a century ago. Ticks that were attached to a cat in Hungary in 1884 and to a fox in Austria in 1888 were collected, marked, and stored but it was not until 1995 that they were actually analyzed. When scientists from the University of Berlin and the Harvard School of Public Health finally ran their tests, they were amazed to discover that the ticks were infected with *Bb.* These proved to be the oldest known record of infection anywhere on Earth. Likewise, white-footed mice collected and preserved by a Massachusetts researcher in 1894 allowed scientists to run DNA tests recently that established the presence of the Lyme bacterium in the United States more than a century ago.

By the 1970s, Lyme disease was beginning to get a minor amount of attention by some physicians in the United States. For the first time, an incidence of EM known with certainty to have been acquired in the United States was reported by Rudolph Scrimenti, who treated a patient who had been bitten by a tick while hunting grouse in Wisconsin.

A great deal of credit for the heightened interest belongs to the partnerships that were forged between parents and professionals. A group of heroic moms and dads were waking up to problems in their communities and beginning to form partnerships with state and local scientists, a collaborative effort that would later play a crucial role developing the critical mass, just waiting for a catalyst to explode Lyme disease into full public awareness. Some parents had the full backing of a supportive family; some did not. Some had enough money to live comfortably; others struggled in near poverty. Some continue their efforts today; others have moved on. A number of their stories are sketched later in this chapter and in the "Timeline" in the appendix, but I can never do full justice to their determination and sacrifices. All of our lives are richer for their work.

Another development was that Allen Steere, who had previously been a CDC epidemiologist, and his colleagues published a now-infamous study in the January/February 1977 edition of *Arthritis and Rheumatism,* reporting an unusual collection of symptoms in 51 patients in the Connecticut towns of Lyme, Old Lyme, and East Haddam. To qualify for the study, patients had to have joint swelling, and most had some other combination of fever, muscle aches and pain, severe headache, vomiting, increased sensitivity to touch, and Bell's palsy. Based on physician and patient reports, the authors suggested the symptoms were linked to some sort of arthropod. Claiming that arthritis had never before been described in relationship to the EM rash, the authors declared they had uncovered a new clinical entity and named it "Lyme arthritis."

They also concluded that antibiotics were ineffective. In a second paper, published in the *Annals of Internal Medicine* in June 1977, they elaborated on that conclusion. While citing 12 European publications that concluded that antibiotics were useful to patients with EM rash and neurologic problems, the authors reported that 25% of their own study patients had developed symptoms despite antibiotic treatment. Although they expressed reservations about other options, the researchers nonetheless treated their patients instead with steroids and aspirin.

Unfortunately, the findings of these pivotal studies were faulty in a number of regards. First, the links between EM and arthritis had already been well described. Second, and more important, we now know that steroids can actually harm Lyme disease patients and that antibiotics—the right ones, at the right time, in the proper dosages—are appropriate treatment. A study comparing antibiotics, perhaps administered intravenously, to other treatment options might well have yielded more useful results and gotten optimal treatment to patients more quickly.

In a sense, Steere and his colleagues were rediscovering a disease that had been known for at least a century, yet the media attention given to their work continues to eclipse 100 years of history. Perhaps one reason that the "Lyme arthritis" and "Lyme, Connecticut" stories receive so much attention is that government and private institutions can share some credit for the findings—the National Institutes of Health (arthritis institute), the CDC, the national Arthritis Foundation and its Connecticut affiliate, Yale University, and the Connecticut Department of Health have all claimed that their help led to the discovery. If the Lyme, Connecticut, stories were put in proper perspective, there would be a lot less credit allocated to one incident. Concern about the risks to children, who initially seemed to be disproportionately afflicted with Lyme arthritis, also helps explain the interest in the problem.

PHASE TWO: BREAKTHROUGH—DISCOVERING THE CAUSE AND CONSOLIDATION OF SYNDROMES

Scientific discoveries are often tremendously serendipitous. A researcher may set out on one quest and end up finding something altogether different. Such was the case when Willy Burgdorfer, an entomologist at the United States Rocky Mountain Laboratories of the National Institutes of Health (and later a founding board member of the Lyme Disease Foundation), began investigating outbreaks of Rocky Mountain spotted fever. Research scientists Jorge Benach of the New York State Department of Health and Edward Bosler, Ph.D., now an entomologist at the School of Medicine at the State University of New York's Stony Brook campus, collaborated in the dogged and dangerous work of gathering and testing ticks for disease-causing pathogens.

Dog ticks were their first target, but a series of tests proved negative. The researchers then shifted their attention to black-legged ticks. Again, there was no indication of the pathogen for Rocky Mountain spotted fever. But in the fall of 1981, one of the batches of ticks yielded something dramatically new.

It began when Burgdorfer noticed an embryonic form of a parasite in the body fluid (hemolymph) of two of the ticks. As a scholar familiar with the European medical literature on tick-borne disorders, Burgdorfer remembered the advice of Austrian dermatologist B. Lipschütz, who had urged scientists to examine the intestinal tract and salivary glands of the tick. He also recalled the writings of Swedish researcher Sven Hellerström, who suggested that a spirochete might be involved. Guided by their theories, Burgdorfer undertook a very close inspection of the tick—and behold, found poorly stained, sluggish spirochetes.

Within a year, the spirochetes had been named *Borrelia burgdorferi* (*Bb*) and definitively identified as the cause of Lyme disease. It was an historic discovery, the moment for which hundreds of researchers had been waiting for more than a century. Burgdorfer, a mentor to countless younger scientists, a man who epitomizes all that is good in public health science, and someone I am honored to consider a good friend, was clearly the man of the hour. His name became one for the history books, and those of us concerned about Lyme disease were positioned at last for an explosion of knowledge.

Next came a period of consolidating and expanding knowledge. Following the discovery of *Bb* and the diseases associated with it, researchers began to learn more about how infection lodges itself in the body. At the Second International Lyme Borreliosis Conference held in Vienna, Austria, in 1985, Paul Duray, a Lyme disease researcher now at the National Institutes of Health in Bethesda, Maryland, as well as a former Lyme Disease Foundation board member, declared that the Lyme disease bacterium disseminates itself through the body early in the course of infection. The prevailing wisdom at the time was that infection was slow to spread, and it took 10 more years before Duray's findings were widely accepted. Also in 1985, Burgdorfer was able to demonstrate that ticks infected with the Lyme spirochete could be found across the country. Unfortunately, prospects for international cooperation were somewhat dampened because of an attitude of superiority on the part of some U.S. scientists. There was also some tacit censorship, with scientific presentations offering only a limited perspective on the disease.

Much more work had to be done. In a number of local communities, the partnerships between parents and professionals were beginning to make headway. Weary of ticks and the diseases they were helping to spread, concerned about friends and family members who were ill, and convinced that local public health officials were slow to acknowledge the problem and to warn community residents, a unique form of scientific partnerships became influential. However, there was no effort to coordinate their work on a national level.

I'd like to introduce you to some of the heroes of the early Lyme disease partnerships.

- For 7 months, **Vicki Korman** was in and out of a New Jersey military hospital after developing a variety of odd symptoms, including panic attacks, swollen joints, difficulty swallowing, and severe headaches. Eventually, after she was diagnosed with Lyme disease and treated appropriately, a physician named **Charles Weber** introduced her to a fellow patient, **Carol**

Gabriel. Together the two women worked tirelessly to raise public awareness about the disorder and to establish New Jersey's first local support group. They also formed a partnership with **Terry Schulze,** an entomologist with the New Jersey Department of Health, to create a public education program about Lyme disease that has been widely distributed to local hospitals. Korman became a founding member of the Lyme Disease Foundation's board of directors, and Schulze has served as one of its scientific advisors, as well as a reviewer for the *Journal of Spirochetal and Tick-borne Diseases,* our peer-reviewed journal.

- **Gloria Wenk** also swung into action from sheer frustration. Although public health officials in her home county of Dutchess, New York, were telling her that ticks did not pose any danger, this community educator knew otherwise. In 1986, she got 1,800 signatures on a petition stating that the county had a tick problem that required action and sent it to the state governor. Soon, she was invited to meet **Edward Bosler,** a research scientist in the New York State Department of Health, and a founding board member of the Lyme Disease Foundation, and began immersing herself in the Lyme disease problem.

Within a year, Wenk had became fondly known as the "Tick Lady." She drove around her county collecting ticks, photographing EM rashes on people, talking to the media, and making presentations about Lyme disease. Wenk simply refused to allow Dutchess County public health officials to close their eyes to the problem exploding around them. She is still fighting the good fight, most recently focusing on tick paralysis, which first appeared in dogs but later left a little girl in the county battling for her life. Her own community has given her a little recognition, but to many of us, Gloria Wenk will always be the hero of Dutchess.

- **Amy Jones** had a firm set to her jaw when she walked into the Fox Chase Cancer Center in Philadelphia in 1988 and said to Paul Duray, then a pathologist at the institution and the first scientist to find *Bb* in human tissue other than skin, "I think we have Lyme disease in my village." Children up and down her block in a Pennsylvania town had fallen ill, but no one was paying attention. Duray accompanied Jones to her house and recalls: "I stepped out of the car and immediately heard the crunching sound of ticks under my shoes. We collected ticks right on the spot." Sure enough, laboratory testing proved they were carrying the Lyme pathogen.

That got Jones moving. She had little formal education, no political contacts, not even a business suit to her name—but that didn't stop her. She cajoled just about every local and state representative she could find into doing something and began an aggressive campaign to raise funds for research and education. Slowly and methodically, Jones got people paying attention to the problem of Lyme disease in Pennsylvania. Every resident and visitor to the state owes her a debt of gratitude.

The work of these people, and that of others unnamed, is an inspiration and a model of how scientists and advocates can collaborate.

PHASE THREE: LYME DISEASE BECOMES A HOUSEHOLD TERM

I don't think I'm being immodest to say that the birth of the Lymé Disease Foundation in 1988, and the struggles of my son Jamie, were the catalyst to heighten the public and scientific awareness, making Lyme disease a household term by 1990. Today, the foundation's focus has been broadened to include all tick-borne disorders, and we remain the largest, most scientifically rigorous, most influential national and international organization dedicated to education, research, and advocacy. Our overall goals after almost a decade of work remain much the same: to arm people with knowledge about how to prevent tick bites and the diseases they can cause; to improve diagnosis and treatment for those who become infected; to push for more wisely used federal funds for research into Lyme disease and other tick-borne disorders; and to bring greater attention to the tremendous public health problem created by infected ticks. We are especially proud of our teamwork with a wide range of internationally renowned scientists, many of whom serve as our scientific advisors.

As the Lyme Disease Foundation drew extensive worldwide media coverage, many other activists were getting busy on the international front. In 1989, **Diane Kindree,** R.N., established the Vancouver, British Columbia, Lyme Borreliosis Society, the first Canadian nonprofit organization dedicated to Lyme disease, working closely with **Satyen Banerjee,** a researcher at the British Columbia Centres for Disease Control. One year later, **Bela Bózsik,** M.D., established the Lyme Borreliosis Foundation of Hungary. Because his organization's funds were so scarce, Bózsik spent several weeks at our Connecticut home when he was attending a scientific conference, and we have continued to support his vital efforts in every way we can. **Terry Moore,** of DeeWhy, Australia, put the problem of Lyme disease in Australia on the map by founding the Tick Alert Group Support (TAGS) in 1993 and forming a partnership with **Bernie Hudson,** a microbiologist at the Royal Hospital in Sydney. At the time, the official word was that the disease did not exist in that country. Three years later, an Australian Ph.D. candidate named **Michelle Wills** proved that it did.

It was the effective partnerships among patients, government officials, and researchers that enabled volunteers around the world to bring Lyme disease the attention that has helped make it a household term. The struggle is waged by other heroes too numerous to mention here. Look around to see if one is living in your community. If so, why not find a way to say "thanks"? A local award ceremony; a special club meeting; a ceremony with town public officials, anytime of the year, or in May, which is Lyme Awareness Month; or a day dedicated to "Volunteers Who Make a Difference" can cost so little and mean so much.

4

SIGNS AND SYMPTOMS OF LYME DISEASE

For almost 100 years, various staging systems have been proposed to describe the pattern and timing of Lyme disease signs and symptoms. The approach I feel is most accurate and useful was proposed in 1990 by Swedish researchers. It views the illness in three distinct stages:

1. Local disease, when the erythema migrans (EM) rash is the only visible sign of disease.
2. Disseminated early disease, characterized by widespread inflammation as the spirochetes begin to spread throughout the body. A range of flu-like symptoms may be present. More severe symptoms at this stage may suggest that more serious and hard-to-treat disease will follow. Researchers recognize that *Bb* sometimes begins to travel through the body within days after a tick bite has occurred, even before an EM rash appears.
3. Disseminated late disease, which results in organ degeneration and damage to many of the body's systems. Late disease usually starts about 3 months after the initial tick bite, but may take much longer to become apparent.

The course of disseminated Lyme disease is far from predictable. New signs and symptoms associated with infection are still being discovered. Clearly, the infection is protean—that is, ever-changing—and because so many body systems are involved, it takes a seemingly infinite array of forms. During periods of active infection, the patient's medical problems may seem totally unpredictable. One day a knee can be swollen. Several days later, the knee will feel fine, but there will be a severe migraine headache. Few people have the full range of disseminated problems, but most people have at least some of them. Some patients with a mild infection show no obvious signs at all, but research suggests that at least in some of them, ongoing tissue damage is nonetheless occurring. In other people, Lyme disease is so severe that it completely disrupts normal life. Lyme disease can also become chronic, or it can follow a cyclical pattern of active infection, remission, and relapse. It also seems likely that patients may be infected with multiple strains of *Bb* following a single tick bite or may be simultaneously infected with other tick-borne disorders, such as human granulocytic ehrlichiosis and babesiosis and may suffer more severe disease as a result.

Symptomatic differences also exist from one region of the world to another. Originally, physicians thought that arthritic conditions were more common in North America while neurological problems occurred more often in Europe, attributing those differences to bacterial strain variations on the two continents. Now, research has proved them wrong. The early impression probably reflected local biases and knowledge gaps that resulted in a failure to recognize the full spectrum of disease. Research by Allen Steere dating back to the 1980s suggested that only about 15% of Lyme disease patients in the United States had neurological problems, but as the signs became more widely recognized, that figure was estimated to be as high as 40%, according to 1993 research by L. Reik at the University of Connecticut. Likewise, Lyme arthritis is now recognized as equally common on both sides of the Atlantic.

In general, the course of Lyme disease in children is quite similar to that of adults, but there are some significant differences. It was recurring arthritic problems that first brought public attention to the disease in Lyme, Connecticut, but researchers now know that chest pain, abdominal pain, headaches, and cranial nerve palsies are the most dominant complaints. Other neurological problems that may occur include light sensitivity, dizziness, stiff neck, numbness or tingling, sleep disturbances, memory problems, generalized or progressive weakness, and more. A neurologic condition called "pseudotumor cerebri-like syndrome," which simulates brain tumors, has also been reported in children. This problem is sometimes observed during an eye examination, because the condition causes swelling of the optic nerve and severe headaches. Our young patients often suffer from sore throats and heart palpitations, and preschoolers are especially prone to irritability and mood changes.

Especially troubling is that disease manifestations in children may easily be mistaken for intentional misbehavior. Because concentration problems are characteristic, Lyme disease may also be wrongly diagnosed as attention deficit disorder. The risk that a child may not receive proper medical treatment or may be wrongly thought to have some sort of psychological or behavioral problem speaks volumes about the need to push scientific research along.

Lyme disease resulting in death has been reported and published in peer-reviewed academic literature. For example, at least four papers describe the deaths of symptomatic patients in whom autopsies revealed *Bb* spirochetes in the heart. There have also been reports in the medical literature of Lyme disease–associated stroke. Nonetheless, there is a reluctance on the part of the scientific establishment to publish about such deaths, or to directly attribute death to Lyme disease, which makes it enormously difficult to know how often death actually results.

The signs and symptoms of Lyme disease described here are extraordinarily extensive. While you are highly unlikely to experience all, or even most, of them, it is important to know what *might* be associated with this infection so that you can seek appropriate medical help promptly.

LOCALIZED DISEASE

Erythema Migrans Rash

The first and most characteristic sign of Lyme disease is a single erythema migrans (EM) rash, which appears at the site of a tick bite and was initially described in 1909. The presence of this distinctive localized rash is sufficient for your doctor to diagnose you with Lyme disease. A mild headache, a slight fever, or a few other aches and pains may accompany the EM rash.

The rash may begin as early as two days after a tick bite, or it may not appear for several months. However, in a significant minority of patients—perhaps as many as one-third or more—it never appears at all or is too light to be noticed. African-Americans and others with dark skin are especially likely to miss the EM rash or to believe that it is a bruise. For a time, children were thought to have the rash less often than adults, but researchers now think it more easily goes unnoticed in children. Figures 4–1 through 4–7 show a variety of EM rashes.

The typical rash begins as a small, reddish bump about one-half inch in diameter that is sometimes raised. It soon begins to expand outward, usually clearing in the center and reaching a final size that may be as little as 2 inches in diameter or may be large enough to cover your entire back or to wrap around your arm or leg until the two ends meet. The characteristic "bull's-eye" form of the rash is a series of bright red and paler skin colors resembling the ripples created when a rock is thrown into a pond. However, there are many other well-recognized variations of the Lyme rash. It may be uniformly discolored and may range in color from reddish to purple to bruised-looking. An EM may be shaped like a ring, triangle, an oval, or a long, thin, ragged

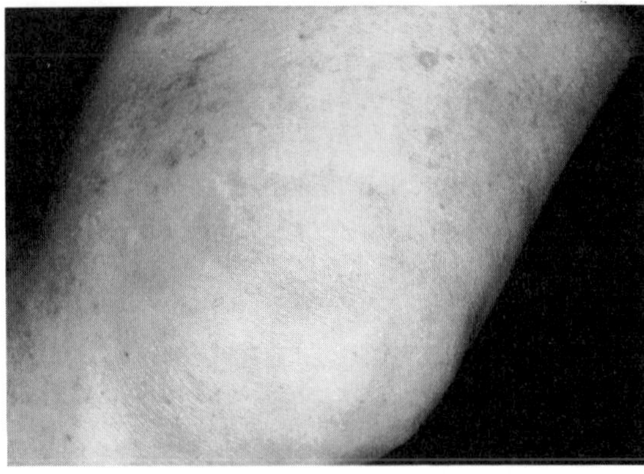

Figure 4–1
Faint EM. Reprinted by permission of LDF.

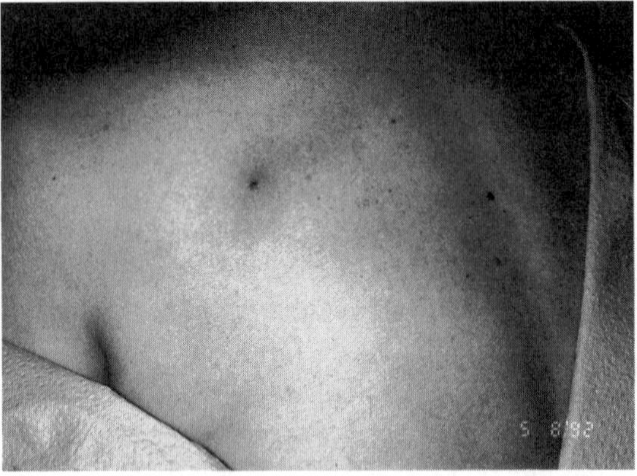

Figure 4–2
Embedded black-legged tick bite with enlarging EM. Reprinted by permission of M. Patmas, M.D.

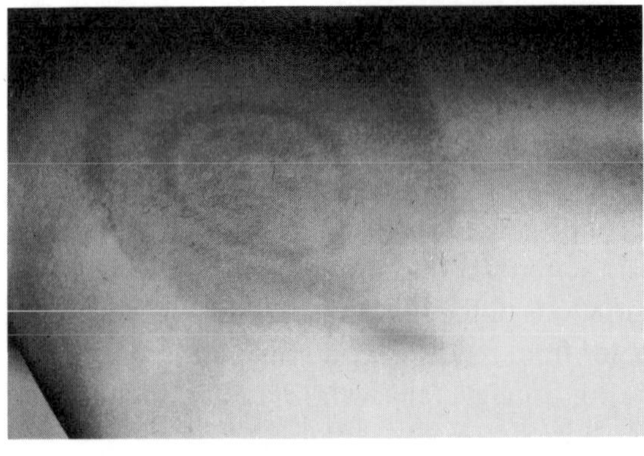

Figure 4–3
Traditional EM. Reprinted by permission of M. Schwartzberg, M.D.

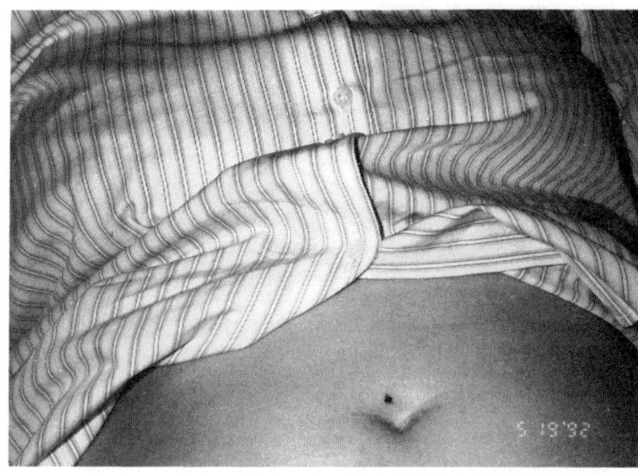

Figure 4–4
Embedded lone star tick with enlarging EM. Reprinted by permission of M. Patmas, M.D.

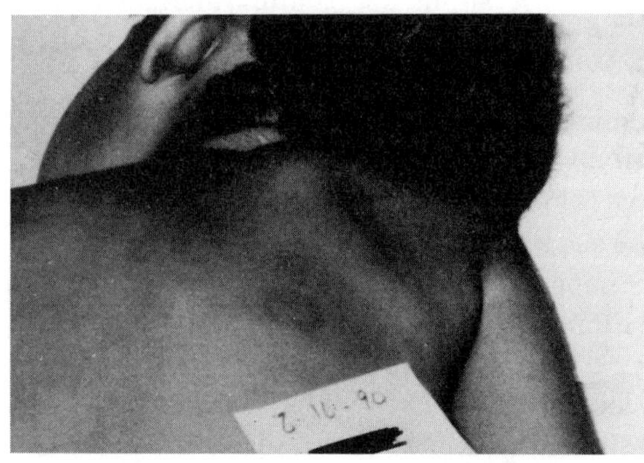

Figure 4–5
EM on dark skin. Reprinted by permission of E. Masters, M.D.

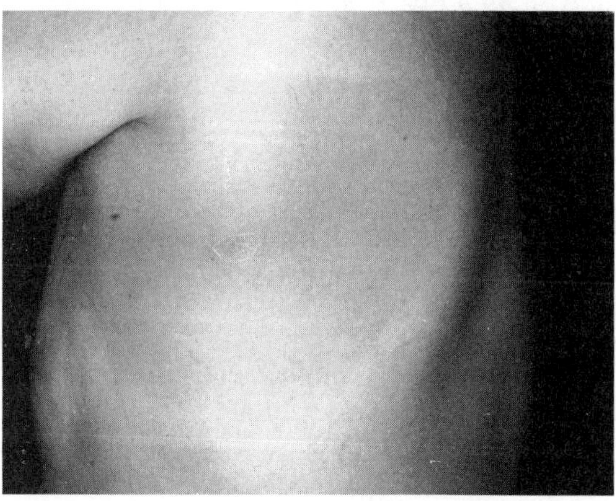

Figure 4–6
Crusting EM. Reprinted by permission of A. MacDonald, M.D.

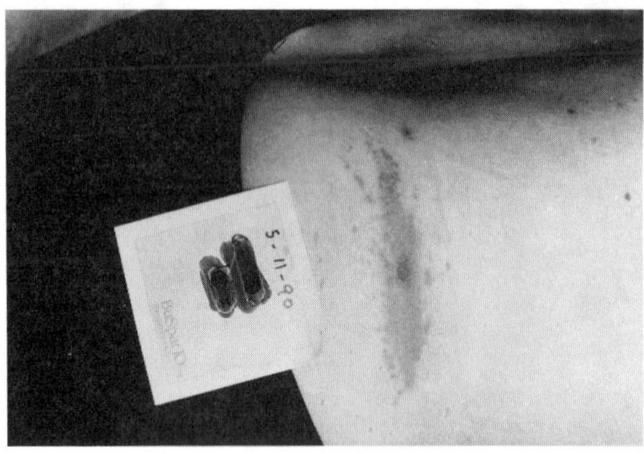

Figure 4–7
EM from Midwest long thin line. Reprinted by permission of E. Masters, M.D.

line and may be hot to the touch, itchy, burning, or painful. Alternatively, the skin at the bite site may be numb. You may also see blistering, ulceration, or scaling, and very occasionally, there will be some bleeding.

Don't confuse a local reaction to a tick bite with a more serious infection. A small inflamed skin bump or discoloration that develops within hours of a bite and then starts to disappear within a day is not likely to be serious. The rash associated with Lyme disease, by contrast, can last from 1 week to several months and eventually disappears on its own, without treatment. Be warned, however: This does not mean that the infection is eradicated. To the contrary, if you do not receive treatment, more serious, systemic problems may begin as the infection becomes disseminated. Unfortunately, infection may have spread when the rash first appeared—one researcher has demonstrated that slightly more than half of all patients in the United States with an EM already have disseminated disease.

If you develop a rash that even *might* be EM, be sure to call your physician. A smart thing to do is to mark the outward edge of the rash with a waterproof marker and take a photograph of it. That way you can easily determine whether the rash has grown.

DISSEMINATED DISEASE

General indications of disseminated infection may include: headache, fatigue, malaise, serious muscle aches and pains, fever, chills, sore throat, profuse sweating, diarrhea, swollen glands, and hoarseness. Certain symptoms point to the involvement of specific organs, such as joint swelling, severe headache, or irregular heartbeats.

Neurological Problems

Neurological problems are now estimated to occur in as many as 40% of Lyme disease patients in the United States. Researchers have been aware of the neurological involvement associated with Lyme disease since 1922, but much more is now understood about how it occurs. Reik, the University of Connecticut researcher, believes that neurological damage may occur in two different ways. First, a direct bacterial invasion may cause damage, degeneration, or wasting of the muscles, nerves, and other body tissue. Second, inflammation of the blood or lymph vessels (vasculitis) may adversely affect the body's ability to nourish organ tissue properly. Interestingly, patients bitten in the head are three times more likely than others to develop neurological Lyme disease. Because children are much more likely than adults to be bitten on the head or neck, they are also more likely to develop neurological conditions.

The three most common neurological conditions are (1) *cranial nerve palsy,* (2) nerve root inflammation, known as *radiculoneuritis;* and (3) *meningitis,* which is an inflammation of the membrane that covers the brain and spinal cord. This is a classic triad of neurological problems known as Bannwarth's syndrome, which sometimes occurs in the early stages of disseminated disease. Each of these conditions will be described in this chapter.

Cranial Nerve Palsy

Twelve pairs of cranial nerves, which branch out from the lower surface of the brain and extend around the body, are each associated with a different brain function. At any stage of disseminated Lyme disease, any of these nerves may lose their ability to conduct electrical impulses properly, leading to *cranial nerve palsy,* commonly called paralysis in areas controlled by that pair of nerves. Cranial nerve palsy is the second most common Lyme neurological condition. Failure of the facial nerves causes Bell's palsy, in which numbness, muscle paralysis, and many other symptoms can occur on one or both sides of the face. Bell's palsy is the most commonly diagnosed cranial nerve abnormality of Lyme disease, followed by optic nerve dysfunction. However, it is unclear whether these actually occur more often or are simply easier to recognize.

Here is what may go wrong with the twelve pairs of cranial nerves, which are typically designated by Roman numerals:

I. Olfactory: There may be a loss of smell, or smells may be overly intense or noxious.

II. Optic: Partial or total loss of vision may occur.

III. Oculomotor: The eyelids may droop, the eyeball may deviate outwards, or the pupils may become dilated. Some patients with a malfunctioning oculomotor nerve may squint involuntarily or see double images.

IV. Trochlear: The eyeball may rotate upwards, and outwards or double vision may occur when looking down.

V. Trigeminal: Pain or numbness in parts of the face, scalp, forehead, temple, jaw, eye, or teeth has been reported. The muscles used for eating may become paralyzed or dysfunctional, making it difficult to chew, and the jaw may deviate toward the paralyzed side.

VI. Abducens: The eye may deviate outwards, and excessive squinting or double vision may also occur.

VII. Facial: The improper functioning of these nerves can result in Bell's palsy on one or both sides of the face. (See Figures 4–8 and 4–9.) Characteristic problems include facial numbness or pain and paralysis of the muscles, sometimes leading to difficulties in chewing or a tendency to dribble food. Patients are unable to wrinkle their foreheads, the lines on the forehead and nose may become abnormally smooth, and the eye droops. An inability to make tears has also been reported. In addition, the jaw deviates to the paralyzed side, and hearing loss can occur on the affected side. Tooth, ear, and jaw pain has been reported. Disruptions of the mucous membrane in the front two-thirds of the tongue result in loss of taste.

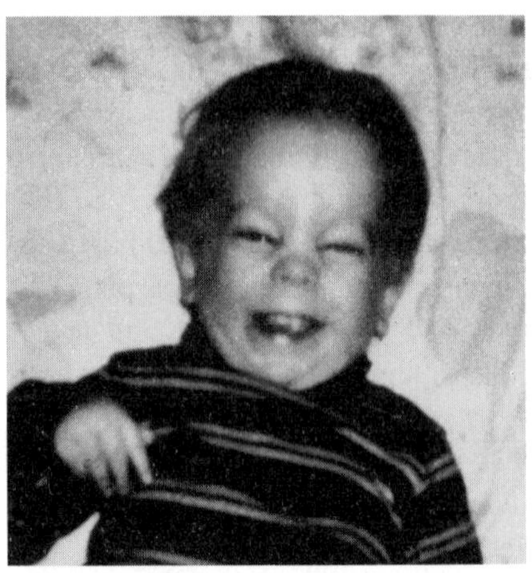

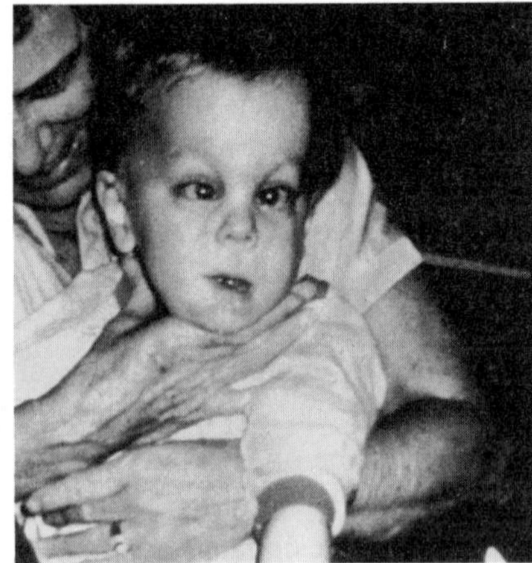

Figures 4–8 and 4–9. My son, Jamie, before and after Bell's palsy. Photos taken one week apart. Reprinted by permission of Karen Vanderhoof-Forschner.

VIII. Vestibulocochlear: Hearing disturbances such as a ringing in the ear or full or partial hearing loss may occur. Equilibrium disturbances such as dizziness, vomiting, or reeling are also associated with nerve dysfunction.

IX. Glossopharyngeal: Problems occur with the mucous membranes in the back one-third of the tongue, leading to abnormal taste sensations, such as excessive sourness or bitterness, in that region. Improper functioning may also result in difficulty swallowing and paralysis of the glottis, a piece of tissue that ordinarily prevents you from inhaling food into your lungs or swallowing air.

X. Vagus: Dysfunction of the muscles in the throat, shoulders, and back may create difficulty in swallowing or talking. Drooping shoulders and an inability to rotate the head away from the dysfunctional side may also occur. Malfunctions within other branches of this nerve may cause heart problems (including palpitations and a rapid pulse); breathing difficulties (including slow respiration and a sense of suffocation); a persistent cough; paralysis of the glottis; vocal cord spasms or paralysis (possibly resulting in an unusually deep voice, hoarseness, or a weak voice); and gastrointestinal disturbances (typically, constant vomiting). These nerves also affect the liver, intestines, spleen, kidney, thyroid, and testes or ovaries.

XI. Spinal Accessory: This nerve pair can be associated with disrupted function or paralysis of the upper back and neck. Back spasms and the inability to tilt the head to the shoulder or to rotate the head in either direction can also occur.

XII. Hypoglossal: One side of the tongue may be partially or fully paralyzed, and the tongue or larynx may deviate toward the paralyzed side, with a resulting slowed articulation, thick speech, and difficulty swallowing.

Other Nerve Involvement

When Lyme disease disrupts the conduction of nerve messages, a wide variety of often-progressive problems can occur, including significant muscle weakness leading to paralysis. Guillain-Barré syndrome, a disease of the peripheral nerves resulting in limb numbness or weakness, can occur. Or the diaphragm may become paralyzed, forcing the patient to be placed on a respirator. A spinal tap, in which a sample of the cerebrospinal fluid that surrounds the brain and spinal cord is withdrawn for testing, may reveal nothing, but nerve conduction tests do sometimes show abnormalities. If the nerve degenerates, permanent damage can result.

Nerve-related symptoms that can occur at any time during the course of disease include:

- Diminished reflexes.
- Sharp, shooting pains that radiate down the arms, legs, or back.
- Areas of numbness, tingling, prickling, or heightened sensitivity.
- Poor muscle coordination, muscle weakness or paralysis, involuntary muscle twitching, progressive muscle weakness, and movement disorders, including abnormal movements of the arms and legs and gait problems.

Two other nerve-related conditions that sometimes occur are radiculoneuritis and polyradiculoneuropathy. *Radiculoneuritis,* the most common Lyme disease neurological condition, is an inflammation of the nerve root, and is generally considered a sign of early dissemination. It can start as extreme radiating pain, abnormal skin sensations, and muscle coordination problems. *Polyradiculoneuropathy* is a nerve dysfunction, but it is not associated with inflammation and is typically considered an indication of later Lyme disease. Polyradiculoneuropathy sometimes develops concurrently with encephalopathy or other late-stage indicators. Pain, tingling, and numbness are common, especially on the limbs, and carpal tunnel syndrome may occur.

The Brain and Spinal Cord

Infection of the brain can cause a wide variety of serious medical problems in both the early and late stages of disseminated disease. About one-third of these patients suffer from depression.

Meningitis: *Meningitis,* the third most common neurological condition, is characterized by inflammation of the membranes covering the brain and spinal cord, signals early neurologic involvement. It can be acute, chronic, or too mild to notice and may wax and wane or continue unabated for months. The most characteristic symptoms are a severe headache or a stiff neck, and there are also reports of light sensitivity, fever, nausea, and vomiting. A magnetic resonance imaging (MRI) scan, which uses principles of magnetism to create images of body organs and tissue, is generally normal, but an electroencephalogram (EEG), used to record electrical activity in the brain, sometimes shows evidence of neurological changes.

Encephalopathy: Encephalopathy is a dysfunction of the brain itself that can occur months to years after initial infection and is considered late-stage disease. About half of Lyme disease patients with brain involvement develop encephalopathy. Cognitive problems—including diminished conceptual ability and lessened capacity to think, learn, and apply logic—are often reported. Mood disturbances such as irritability, bursts of crying, and temper flares occur, and patients also report poor concentration, short-term memory loss, profound fatigue, confusion, and difficulty sleeping. Most neurological examinations are normal, although an MRI scan sometimes shows multiple sclerosis-like white matter lesions on the brain caused by inflammation. The most accurate way to diagnosis encephalopathy is usually through neuropsychiatric testing.

Encephalomyelitis: Encephalomyelitis is an inflammation of the brain and spinal cord, which generally occurs in the later stages of disseminated disease, beginning from 1 month to several years after an infection occurs. Symptoms may be difficult to distinguish from those seen in multiple sclerosis, stroke, or brain tumor. The condition is progressive and may suddenly worsen, followed by partial improvement and then another attack. Movement is impaired, as is some cognitive ability. A patient may appear to be dyslexic when speaking or writing, and personality changes are sometimes evident. A range of other alarming symptoms also occurs, including the inability to feel sensations, paralysis of one or both sides of the body, defective muscle movements, bladder dysfunction, visual and hearing loss, seizures, cranial nerve paralysis, and abnormal tissue growth. MRI scans often show excess fluid, tissue atrophy, and white-matter lesions. Spinal fluid tests show abnormalities quite often, and mild brain abnormalities have been found at autopsy.

Sometimes meningitis is concurrent with encephalomyelitis—the resulting condition is called meningoencephalmyelitis.

Psychiatric Syndromes

Psychiatric manifestations of Lyme disease have been reported since as far back as 1930. More recently, researchers have noted that psychiatric disorders appear to be characteristic of Lyme disease at any stage of disseminated infection. Depression is common in as many as one-third of all patients. Infection has also been associated with mood changes, hallucinations, eating disorders, and marked personality changes. Other reported psychiatric syndromes include panic attacks, disorientation, extreme agitation, impulsive violence, verbal aggressiveness, inappropriate, manic, or obsessive behavior, loss of contact with reality, paranoia, schizophrenic-like states, dementia, and psychosis, among others. Tragically, some patients are tormented by suicidal thoughts, and there have been a number of suicide attempts.

An increasing number of patients are having their Lyme disease first diagnosed by a psychiatrist. I expect that in the next 5 years, we will hear a lot more about the toll being taken by Lyme disease–associated psychiatric problems. Fortunately, antibiotic treatment seems to help at least some patients with psychiatric symptoms.

Ophthalmologic Problems

There is evidence that the Lyme spirochete rapidly invades both eyes soon after infection occurs, but it may remain undetected for a long period of time, possibly causing slow damage all the while. Conjunctivitis, an infection of the mucous membrane inside the eyelid that is often called "pinkeye," is typically the first observable eye symptom and may appear months or even years after a tick bite. Inflammation of all parts of the eye has also been reported. Among the other eye problems that sometimes occur are loss of vision, blurred sight, decreased perception of light or color, a drooping eyelid, or annoying floaters or spots appearing in the line of sight. Certain eye problems, such as double vision, wandering or lazy eye, and extreme light sensitivity, may also be signs of neurological involvement linked to malfunctions in the cranial nerves.

Unfortunately, ophthalmologists are not always alert to the possibility that subsequent eye infections may be associated with Lyme disease. The failure to attribute ophthalmologic problems to *Bb* can also result from the relatively high number of false negative antibody tests for Lyme disease in eye patients.

Cardiovascular Problems

The association between Lyme disease and heart problems has been recognized since 1934. While heart problems sometimes resolve spontaneously, they have also been associated with several deaths. In at least four autopsy reports, *Bb* spirochetes have been found in the heart.

Lyme Carditis

Somewhere between 10% and 20% of infected patients suffer symptoms of cardiac infection, which may begin within days, months, or years after the infected tick bite and is considered a sign of early bacterial dissemination. Disease may persist or resolve on its own. Even if the symptoms do resolve, however, the infection may not be eradicated, instead lingering to cause undetected, but sometimes permanent, heart damage. The risk exists that antibiotics prescribed for other reasons may help resolve heart problems without eradicating the bacteria altogether, thus leaving someone with chronic but rarely detected disease.

Lyme carditis can slow the heart down sufficiently for it to require a temporary pacemaker. Alternatively, the heartbeat may become fast or irregular, or palpitations and extra heartbeats can occur. Inflammation can also cause swelling and an enlarged heart. Less disabling but nonetheless alarming symptoms include fatigue, dizziness or lightheadedness, diminished exercise tolerance, chest discomfort, and shortness of breath. At times, chest pain may become more intense, and a patient may go into convulsions or become unconscious.

Some degree of heart block, in which the ventricle chambers beat independently from the atrial chambers, resulting in decreased cardiac output, occurs in as many as half of all Lyme disease patients. On extremely rare occasions, congestive heart failure can occur. Other cardiovascular problems, including palpitations, rapid pulse, and stroke, appear to have their origins in the neurological problems described earlier.

Blood Vessels

An inflammation of the blood vessels occurs during early and late dissemination, constricting blood flow in some patients. There have also been 11 published reports of strokes and some reported heart attacks in Lyme disease patients.

Skin Problems

Erythema Migrans

When a single erythema migrans (EM) is accompanied by severe flu-like symptoms or evidence of other organ involvement, it is considered early disseminated disease. Sometimes, multiple EMs are present, a sign that spirochetes have traveled to another area to create a satellite station. The secondary rash is usually smaller than the primary rash at the bite site. Multiple EMs in response to a single bite occur in about 17% of U.S. patients. (See Figure 4–10.)

Borrelia Lymphocytoma

Borrelia lymphocytoma (BL), first reported in 1911, is a bluish-red, pimple-like lump in the skin filled with white blood cells and *Bb*. If present, this sign of disseminated early disease becomes apparent within a few weeks or months

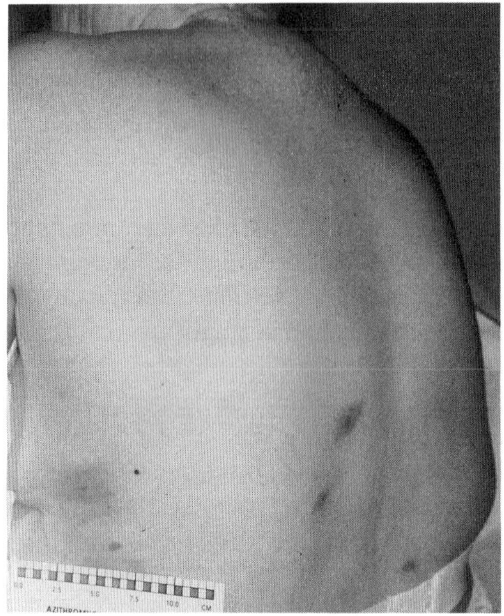

Figure 4–10
Multiple EMs. Reprinted by permission of E. Masters, M.D.

after a tick bite. The lump is generally visibly inflamed unless it develops underneath the skin, in which case you may not be able to see it. In children, a single, localized BL is most commonly located on the earlobe. In adults, a single BL typically appears on one of the nipples of the breast, but it may also appear on the nose or scrotum. Multiple BLs can occur and if they do, may appear anywhere on the body.

BL is more frequently reported in Europe than in the United States, but we don't know why. Possibly it is because U.S. physicians have not been educated about the condition. It may also reflect strain variations, genetic differences in the population, or the depth of the tick bite.

Acrodermatitis Chronica Atrophicans

Acrodermatitis chronica atrophicans (ACA) is the first manifestation of Lyme disease to have been described, with reports dating back to 1883. ACA is a degenerative and chronic skin condition that leads to atrophy of the skin, muscles, and joints. (See Figure 4–11.) A sign of late-stage dissemination, it may begin anywhere from 6 months to 8 years after a primary erythema migrans rash and progress over a period of years. Curiously, if a patient has received antibiotics for some other condition, this rash may not be cured but often looks markedly different.

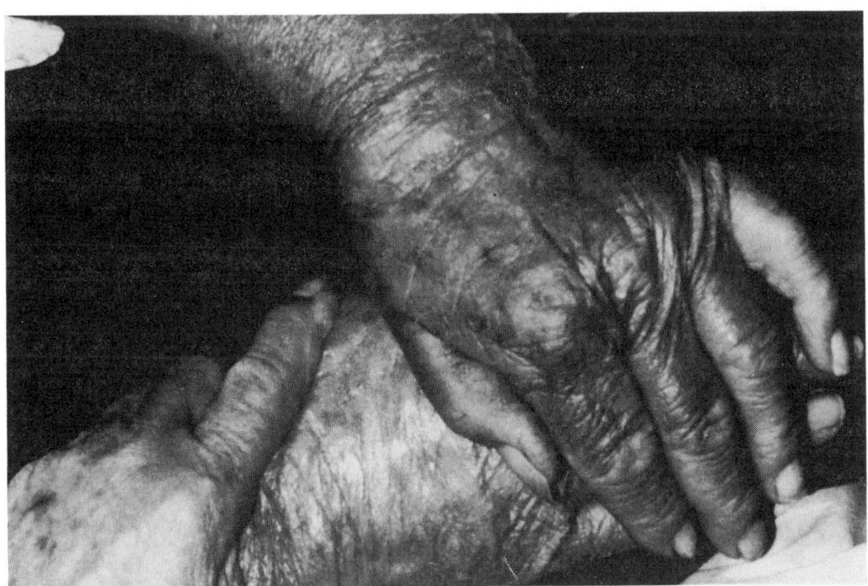

Figure 4–11. Acrodermatitis chronica atrophicans (ACA). Reprinted by permission of E. Masters, M.D.

Typically, ACA starts as a bluish-red patch of discolored skin, usually on the knee, the bottom of the foot, the elbow, the back of the hands, or the buttocks. Over time, ACA changes the character of the skin dramatically. (See Figure 4–12.) Extreme swelling may occur, and the skin becomes reddish-white, mottled, and doughy. As subcutaneous fat is lost, the skin sometimes becomes wrinkled, thin, scaling, dry, and transparent. Alternatively, it may become yellowish, shiny, and parchment-like. Eventually, the skin may lose all color or become excessively pigmented. Loss of tissue in the mucous membranes of the nose, throat, tongue, and vagina may occur. Hair loss, an increased sensitivity to sensory stimulation (such as touch, sound, or taste), or the development of large knotty bumps on the elbows, knees, or finger joints may also occur. There is some evidence that ACA-infected skin tissue sometimes develops into a form of skin cancer.

This skin condition provides important insights into the process of prolonged latency and chronic infection that characterizes Lyme disease. Indeed, Rudolph Scrimenti of the Medical College of Wisconsin has said "it may serve as an ideal study model for persistent bacterial infection." ACA is fairly common in Europe but is rarely diagnosed in the United States, possibly because it is not well recognized.

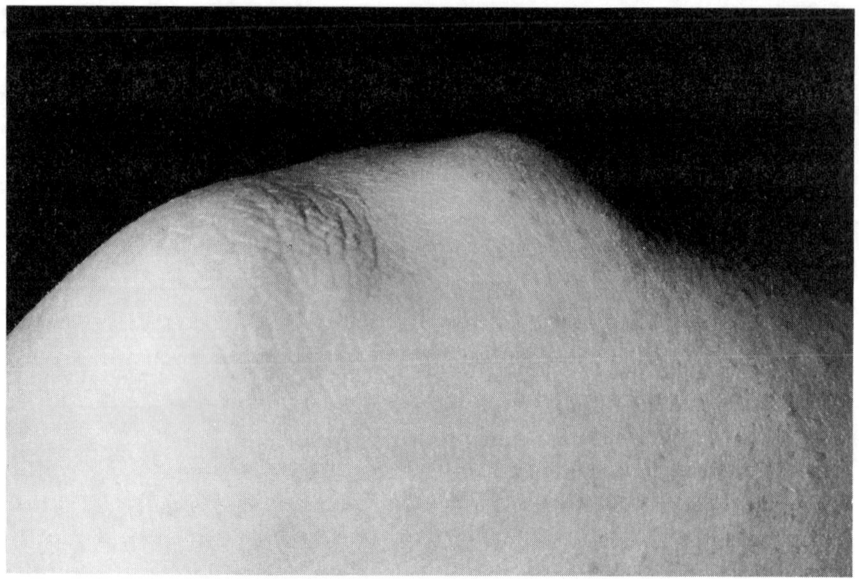

Figure 4–12. ACA fibrous nodule. Reprinted by permission of F. Strle, M.D.

Joint and Muscle Problems

Joint Involvement

Reports of arthritic conditions associated with Lyme disease first appeared in 1921. Lyme arthritis, a sign of both early- and late-stage dissemination, can begin within days or months after a tick bite. Joints and tendons are often warm to the touch and may be painful, swollen, or stiff. The knees are most frequently involved. Other common targets for arthritis are the shoulders, ankles, and elbows. Inflammation in the temporomandibular joint can cause jaw pain severe enough for a patient to consult a dentist. Lyme arthritis is rarely symmetrical—at any given moment, either knee may be affected while the other is not. The condition may start and stop intermittently, with pain lasting for hours or days, interspersed with periods of relief that last several weeks or months.

In late-stage dissemination, tendons, ligaments, and the synovial membrane, which encloses the joints, can thicken, calcify, or change into bone. Some patients have destructive joint changes, including the loss of bone and cartilage.

Muscle Involvement

During early-stage disseminated disease, the fascia membrane, which covers, supports, and separates the muscles, may become inflamed, resulting in pain, stiffness, and swelling. Late-stage disease may lead to a hardening of muscle tissue. Severe cramping, particularly in the calves, thighs, and back, and a loss of muscle tone have also been reported.

The muscles themselves may also become inflamed in early disease, causing swelling and pain, especially in the calves, thighs, and back. Some patients also report severe muscle cramping and loss of muscle tone. Late-stage muscle involvement can result in muscle wasting.

Other Organ Problems

Numerous other organs and body systems may be affected by Lyme disease. Infection that lodges in the liver can cause problems that range in severity from slightly elevated enzymes to hepatitis. Lung problems include breathing difficulties and pneumonia. Gastrointestinal infection may result in nausea, vomiting, diarrhea, reflux, and loss of appetite. An enlarged or tender spleen is sometimes reported. Some patients have had underactive thyroid glands, which return to normal after treatment. Bladder and kidney problems have been reported. There have also been reports of infertility problems.

5

THE QUEST FOR A DIAGNOSIS

Despite the breadth of symptoms that characterize Lyme disease, it remains an extremely difficult disease to diagnose. Although multiple laboratory tests are available, it is impossible to be entirely sure what the results mean. If the laboratory reports are negative, it is no guarantee that you do not have Lyme disease—all you are likely to know with certainty is that the test did not find antibodies to the *Bb* spirochete in your system. No test can definitively rule out Lyme disease. By the same token, positive results do not necessarily indicate that you currently have Lyme disease—it may mean you were infected with *Bb* in the past and created antibodies to fight it.

The struggle to secure an appropriate diagnosis takes an emotional and physical toll and can be enormously costly. In our actuarial research study, published in the January/February 1993 issue of *Contingencies,* we calculated the average cost of treating Lyme disease to be about $62,000 per person and showed that costs are directly related to the time it takes to obtain a diagnosis:

Length of Time to Diagnosis	Average Cost
Less than 6 months	$34,000
7 to 12 months	$68,000
More than 12 months	$92,000

Significantly, half the costs are for medical bills incurred *before* the diagnosis is actually made. But the expense associated with the search for a diagnosis is just one reason why developing a perfect test for Lyme disease should be a top research priority. Until we know who is actively infected and when the Lyme disease bacteria have been eradicated altogether from an individual's body, it is impossible to determine optimal treatments. In fact, without better diagnostic techniques, we can't even measure the true extent of the public health crisis we are facing.

In my years at the Lyme Disease Foundation, countless press releases have crossed my desk from private and academic laboratories, each praising a new, and supposedly highly accurate, test to detect *Bb*. A year or two later, another press release arrives from the same laboratory or academic institution announcing still another "improved" test. It makes me suspect that the earlier versions weren't quite as accurate as the laboratory claimed—and makes me wonder just how good the latest "breakthrough" really is. Unfortunately, many physicians base their clinical decisions on these imperfect tests, which means that many patients being evaluated for Lyme disease infection are not being properly diagnosed or treated.

Having said that, I don't want to discredit the available lab tests altogether. They can be very valuable for helping your physician confirm a Lyme disease diagnosis, especially if the typical "bull's-eye" rash is missing. They may suggest the need to search for other causes of a medical problem. The lab tests can also be useful for monitoring your progress during treatment. It is prudent, however, to recognize their limitations and vital to keep working with scientists for improvements.

GETTING A DIAGNOSIS

Whether or not you remember being bitten by a tick, it is critical that you contact your physician if an unusual rash develops on your body. If for any reason you cannot get to your doctor's office right away, take the advice I offered earlier and mark the outer edge of the discolored rash area with a pen. Take a photograph immediately and again two or three days later so that you can document the rash's expansion. An erythema migrans (EM) rash is the only sign that by itself allows your doctor to diagnose Lyme disease with certainty.

If you do not have a characteristic EM rash but have some of the other signs and symptoms associated with Lyme disease, you and your doctor will have more complicated work to do. These are some of the questions to think about on your search for a diagnosis:

- **Do you have signs or symptoms that indicate Lyme disease?**

In the absence of a definitive laboratory test, Lyme disease remains a clinical diagnosis that is primarily based on your signs and symptoms. Your physician will take a complete history and ask you about rashes and any new or unusual symptoms you have experienced. Many of the skin, musculoskeletal, neurological, and cardiac manifestations of Lyme disease that I described in the last chapter can also accompany other medical conditions, so your physician will most likely conduct tests to rule these out.

- **What do your Lyme disease tests suggest?**

A number of different tests are available to identify *Bb* antibodies. These tests, described later in this chapter, are used, either singly or in combination, to support a Lyme disease diagnosis. Tests are being developed to detect the bacteria directly, but there is no research into tests that could declare a patient bacteria-free. While helpful, results from the tests currently available can also be confusing and may even provide erroneous information. For example, some Lyme disease patients with a persisting infection may be so seriously infected with *Bb* that they are unable to produce antibodies to it, and therefore will test negative on the *Bb* antibody test.

Your physician may also recommend other tests, including nerve conduction and magnetic resonance imaging, which are sometimes used to assess neurologic signs and symptoms. These do not allow a definitive Lyme disease diagnosis but may provide supporting data.

- **Do your non-Lyme disease–related tests indicate something else?**

Other diagnostic tests may allow your physician to identify other medical conditions that could be causing some or all of your problems. However, the possibility that you also have Lyme disease should not be automatically dismissed. Be sure to ask your doctor about the prospect of having two illnesses concurrently.

- **Have you been exposed to ticks that transmit Lyme disease?**

The local tick infection rate in your hometown or at recent vacation sites may suggest the probability of contracting this infection, although it is unlikely to give you definitive information. Unfortunately, many places don't maintain accurate or current infection rates, or you may have difficulty learning what they are, so it is often hard to answer this question.

ANTIBODY TESTS TO DETECT *Bb*

As I have suggested, most laboratory tests look for antibodies to the Lyme disease bacteria, rather than the bacterium itself. If your doctor suspects Lyme disease, an antibody test is usually the first test ordered. Antibody tests are generally run on blood serum, but almost any bodily fluid can be used, including spinal fluid, breast milk, joint fluid, and urine.

An ideal antibody test is specific enough to measure the presence of antibodies only to the Lyme disease bacterium and is sensitive enough to detect everyone who has them. As I have emphasized, however, the tests are imperfect, and for now, the more specific the test, the less sensitive it seems to be, and vice versa. Current research is focused on eliminating "false positives," results that identify people as having Lyme disease when they do not. While important, the trade-off in the scientific design is that an increasing number of people with Lyme disease may be missed.

How an Antibody Test Is Conducted

To test your blood for antibodies, a sample is drawn into a tube and then spun in a centrifuge to separate the dark red whole cells from the straw-colored fluid called serum, which contains the antibodies. Antibodies are classified as one of five different immunoglobulins (Ig), each of which has a different function. If Lyme disease is suspected, your physician may order a test to measure IgM and IgG levels. These are the five immunoglobulins:

1. IgM is the first antibody produced in response to almost any foreign invasion, generally within about 2 weeks of an initial infection. IgM levels peak at about 4 weeks and then decline. Although a positive IgM test is thought to suggest a relatively new infection, some patients maintain IgM antibodies for many years, which may be a bad prognostic sign. A mother's IgM antibodies do not cross the placenta, so it is the test of choice to determine if a newborn has Lyme disease.
2. IgG is the primary immunoglobulin produced to combat toxins, viruses, and bacteria. The body starts producing IgG several weeks after an initial infection, with peak production occurring at about 6 to 8 weeks after infection, followed by a decline. IgG antibodies cross the placenta to protect the baby, so that if the mother has produced antibodies to Lyme disease, the newborn will test positive without necessarily being infected with *Bb*.
3. The IgA immunoglobulin is generated to protect your mucous membranes, repiratory tract, intestinal region, saliva, and tears.
4. IgE is also produced by the respiratory and intestinal tracts, especially among people with allergies.
5. IgD is present in serum in very small amounts, but its function is not known.

Test Results

Interpreting the results of a Lyme disease antibody test is tricky and controversial because *no* test can prove that *Bb* has been eradicated completely. There is also a great deal of controversy about what the results actually mean.

Some scientists believe that people who have had Lyme disease will always test positive for *Bb* antibodies, even if the bacteria themselves have been eradicated. Other physicians believe that as a patient improves, the antibody test will gradually change from positive to negative. Some patients seem to be infected with Lyme disease bacteria yet test negative for the antibodies; others initially test negative and then convert to positive once treatment starts.

Despite limited knowledge about what the findings mean, most laboratories report the results of a Lyme disease antibody test as positive, negative, or equivocal. However, different laboratories may report different results from the same blood sample. The same lab may even report different results from two samples drawn simultaneously but run at different times.

Be sure to request a copy of your lab results for your permanent medical records. At a minimum, you should know where the test was conducted, how you scored, and what cutoffs the lab uses to interpret test results.

Testing Positive

A positive test means that the laboratory has detected measurable levels of antibodies to the Lyme disease bacteria in your blood sample. The significance of this finding is not always clear. Some physicians believe false positive results remain a problem, but I believe they are now rare. Over the years, testing has become more standardized, and the level of antibodies required to report positive results has been raised. In addition, a second test is now required to confirm preliminary results. Nonetheless, false positives might occur if the laboratory makes a mistake or because of a cross-reaction, which occurs when the Lyme test mistakenly measures other antibodies in your system as *Bb* antibodies.

Testing Negative

A negative test does not eliminate the possibility that you have Lyme disease. It just means that the laboratory did not find measurable Lyme disease antibodies in your sample. Although scientific understanding of how *Bb* interacts with the human immune system is incomplete, new findings are shedding light on why infection is sometimes missed, resulting in a "false negative." These are some of the possible reasons:

- Your sample was taken too soon after infection for your immune system to have mounted a detectable response. An antibody response is rarely measurable when the EM rash first appears. The only good reason to be tested at this stage is to compare the "before" level with a sample taken about 6 weeks later.
- You are producing a detectable level of antibodies, but the laboratory made a mistake.
- You are producing antibodies to a strain of *Bb* that the laboratory cannot detect.
- You are producing antibodies, but they are bound to the Lyme disease bacterium (creating what is called a complexed antibody) without enough that are free-floating and detectable.
- Your immune system is compromised and is not responding properly to the bacterial invasion.
- By taking antibiotics early in the disease, you have aborted your immune system response. This is not necessarily bad because the immune response does not always eradicate *Bb* anyway.
- The bacterium has changed its make-up, and your immune system has not noticed it.
- The bacterium is cloaking itself within an immune system cell, limiting the body's capacity to identify and combat it. That discovery that the Lyme bacterium breaks out of the cell it has invaded and causes the cell membrane to collapse and shroud it is very new and potentially significant and will have to be researched further.

Equivocal Test Results

An equivocal reading means either that antibodies are present, but at a level that is not definitely considered either positive or negative, or that the identified antibodies are not exclusive to *Bb*. Repeat testing should be done when the results are equivocal.

TYPES OF ANTIBODY TESTS

A number of different tests for *Bb* are in use and may be run alone or conducted consecutively to confirm prior results.

Enzyme-Linked Immunosorbent Assay

The Enzyme-Linked Immunosorbent Assay (ELISA) test is automated, ensuring that the same standards are applied each time the test is run and allowing a positive sample to be automatically reexamined. The automation also makes it possible to run many samples at once.

In an ELISA test, enzymes are used to detect *Bb* antibodies. Pieces of the Lyme bacteria become stuck to the bottom of a test tube containing the blood serum sample. Any free-floating antibodies will attach and bind to the stuck bacteria, after which the test tube is washed, leaving the bound bacteria and antibody intact. Additional agents are added to the solution, and if it turns yellow or blue, it is considered positive for Lyme disease, with more intense colors resulting in higher readings. Clear or white wells are reported as a negative test.

I recommend the ELISA "capture test" that is available in some laboratories because it uses two different methods to collect results and is highly sensitive. In essence, this test captures all of one type of immunoglobulin being produced in your body, regardless of the reason, and compares it to the amount of antibody being produced specifically against Lyme disease bacteria.

A positive ELISA test is usually confirmed with a Western blot test.

Indirect Fluorescent Antibody Test

The Indirect Fluorescent Antibody (IFA) test was the first test available for Lyme disease. It is labor-intensive and prone to human error. Nonetheless, it is still being used in some laboratories.

In order to run the IFA test, a diluted human blood serum sample is added to a slide containing whole dead *Bb* bacteria. Free-floating antibodies then bind to the bacteria. Reagents and fluorescent dye are applied to the sample, which is placed under a microscope so that a technician can look for a reflective green color that is visible under an ultraviolet or fluorescent light. A green color shows the binding of animal antibodies to the human blood/antibody material and indicates a positive test.

A positive IFA test should be confirmed with a Western blot test.

Western Blot Test

A Western blot test is usually conducted after a positive ELISA or IFA test and uses the same blood serum sample. In the Western blot, detergent is used

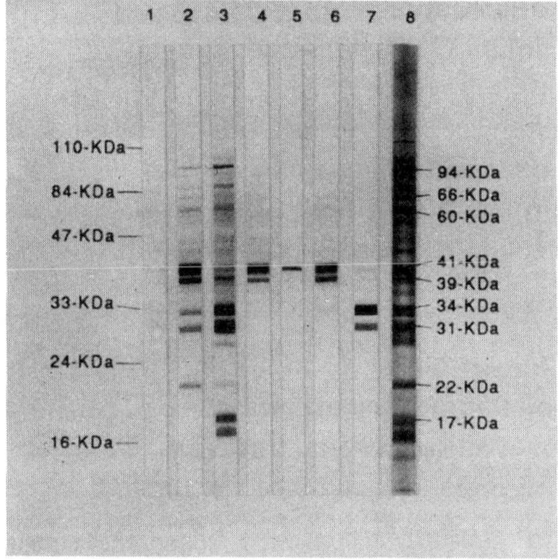

Figure 5–1
Western blot sample. Reprinted by permission of N. Harris/Igene-X, Inc. Reference Lab.

to break one strand of *Bb* into little pieces. This mixture is then placed at one side of a gelatin-like strip and charged with an electrical current, which causes the various pieces of *Bb* to separate by weight. The lighter pieces move more quickly to one end. A very thin sheet of paper is placed on top, and the serum sample is added over the paper. Any bacterium-specific antibodies then bind to the corresponding bacterium pieces below. A reagent is added that causes a color change when antibodies are present.

In the Western blot test, researchers look for a series of bands that resemble rungs on a ladder. (See Figure 5–1.) Each band represents an antibody that is being produced in response to various pieces of the Lyme bacterium. The more bands, the more pieces of the Lyme bacterium to which you are reacting and the longer you have been infected. If you keep getting more bands, the likelihood is that you are getting sicker. Make sure that your physician requests that all reactive bands be reported to assure the most comprehensive information possible.

Many people are given a copy of their test results and try to figure out what the bands mean. These guidelines should help, although you may want to discuss the results with your physician:

- 9kD = cross-reactive for *Borrelia*
- 12kD = specific for *Bb*
- 18kD = unknown
- 20kD = cross-reactive for *Borrrelia*
- 21kD = unknown

- 22kD = specific for *Bb*, probably really the 23/25 band
- 23/25kD = outer surface protein C (OspC), specific for *Bb*
- 28kD = unknown
- 30kD = unknown, probably an outer surface protein; common in European and one Californian strain
- 31kD = outer surface protein A (OspA) specific for *Bb*
- 34kD = outer surface protein B (OspB) specific for *Bb*
- 35kD = specific for *Bb*
- 37kD = specific for *Bb*
- 38kD = cross-reactive for *Bb*
- 39kD = major protein of *Bb* flagellin; specific for *Bb*
- 41kD = flagellin protein of all spirochetes; this is usually the first to appear after *Bb* infection and is specific for all *Borrelia*
- 45kD = cross-reactive for all *Borrelia*
- 50kD = cross-reactive for all *Borrelia*
- 55kD = cross-reactive for all *Borrelia*
- 57kD = cross-reactive for all *Borrelia*
- 58kD = unknown, but may be a heat-shock *Bb* protein
- 60kD = cross-reactive for all *Borrelia*
- 66kD = cross-reactive for all *Borrelia*, common in all bacteria
- 83kD = specific antigen for the Lyme bacterium, probably a cytoplasmic membrane
- 93kD = unknown, probably the same protein as 83, just measured incorrectly

A 1994 CDC grant to the Association of State and Territorial Public Health Laboratory Directors developed the following very controversial guidelines for what is considered a positive Western blot. Some states have quietly implemented these guidelines in the laboratories that do business in their states. To qualify for recent infection, a patient needs two of the three following *Bb* IgM bands: 23/25kD, 39kD, 41kD. To qualify for longer-standing infection, a patient needs five of the ten following *Bb* IgG bands: 18kD, 21kD, 28kD, 30kD, 39kD, 41kD, 45kD, 58kD, 66kD, 93kD. All these are based on strains B31, 297, 2591. Other strains may make other bands.

A number of scientists do not consider these criteria adequate because they were developed from studies of a single strain of *Bb,* a strain that the Association of Laboratory Directors itself advised against using in test development. Serious controversy exists about which bands in a Western blot test are most significant and how many are needed to allow a diagnosis of Lyme disease. Some scientists have proposed alternative criteria for a positive IgG

Western blot that would require any two of the following bands to be reactive: 20, 23, 35, 39, or 88kD. Others believe that having a single band specific for Lyme disease may be sufficient for a diagnosis. Some critics have also objected to dropping measurements of OspA and OspB. A further problem is the difficulty that sometimes arises in determining whether faint lines are actually bands.

Obviously, a lot more work needs to be done to increase the value of this confirming test.

Borreliacidal Antibody Test (Immobilization Test)

The Borreliacidal antibody test is actually a new application of the syphilis immobilization test. In order to detect Lyme disease, the Borreliacidal antibody test combines *Bb* with a sample of your blood serum. A technician then looks through a microscope to see if the *Bb* have been immobilized. If so, it is assumed that your immune system has produced antibodies to kill the Lyme bacteria. The test appears to be free of cross-reactions and is best used in patients with late Lyme disease, who are most likely to be generating productive antibodies.

This test was first used to detect Lyme disease by Charles Pavia at New York Medical College and adapted for human beings by Steve Callister at the Gunderson Medical Foundation in Wisconsin. The reaction is strain-specific, requiring the patient to be infected with the same *Bb* strain that is used for the test.

Unfortunately, the Borreliacidal antibody test has several drawbacks. For one, it requires the patient to be off antibiotics for several weeks to ensure that human antibodies in the blood sample, rather than the medication, are responsible for killing the bacteria. Test results are also predicated on the assumption that immobilizing the bacteria has killed them. Another limitation is that the test takes weeks to process. Of uncertain significance is that results often differ from the findings of the ELISA or the IFA tests.

Immune Complex Test

In a responsive immune system, human antibodies may attach, or bind, with a bacterial agent or some other foreign substance to create an immune complex. The immune complex test first unbinds the immune complexed antibodies, turning them into free-floating antibodies. Antibodies in the sample can be measured by some of the other diagnostic tests described here.

DIRECT DETECTION TESTS

Eventually, a definitive test will be developed to detect the presence of the Lyme disease bacteria, rather than antibodies to them. For now, however, direct detection is still being refined. It is also labor-intensive, expensive, and complex to conduct, and has not yielded very impressive or usable results. Nonetheless, I am reviewing some of the research avenues being pursued because a sensitive detection tool would be an enormous boon to public health and could make a big difference in a patient's long-term medical care.

Culturing

The ideal way to make a definitive diagnosis of Lyme disease is to culture *Bb* from tissue or body fluids. This involves placing a sample, such as tissue, blood serum, or spinal or joint fluid, in a vial with a medium that allows the sample to grow. After an incubation period of weeks or months, the *Bb* bacterium will reproduce if it is present, allowing the specific strain to be identified. Some patients have had as many as four different strains of *Bb* cultured from a single sample.

Unfortunately, few laboratories have the expertise to grow the bacterial culture, and *Bb* is notoriously slow to reproduce, delaying test results for a long time. Taking a skin biopsy from the leading edge of an EM rash is the best way to obtain a culture, but even this method has only a 30% to 50% success rate. Nonetheless, if you have an atypical rash or other unusual symptoms, or if you are having a tissue or fluid sample taken for another purpose—for example, a spinal tap—you may want to discuss with your doctor the benefits of securing a separate sample for a culture. Find out first what costs will be covered by your insurance, and make sure the laboratory your doctor chooses is experienced in culturing *Bb*.

Staining

Another direct detection method is to biopsy skin or other human tissue and send it to a laboratory to be stained and examined for *Bb*. The most reliable source of *Bb* is at the outermost edge of the rash, but even there the technician may not be able to find any bacteria. Staining is time-consuming and often fails to provide any useful information, but when it does, the results can be extremely valuable.

Polymerase Chain Reaction Test

The polymerase chain reaction (PCR) test repeatedly replicates, or amplifies, the DNA in a sample. Traditionally, this amplification technique is done once to amplify a single strain of the bacterium. But more recently, laboratories have begun to do "nested" PCR tests, which produce better results. In nested testing, the lab does a second PCR amplification of a small portion of the first sample, using a different bacterial strain or a different set of bacterial pieces. Unfortunately, this process is highly susceptible to contamination.

A PCR test can generate either false positive or false negative results. False positives occur when the sample is contaminated by *Bb* DNA in the laboratory, especially by inadequately cleaned test tubes. False negatives may be reported if there are too few *Bb* to amplify, or if your strain of infection differs from the strain on which the test is based. False negatives can also be a problem if your blood has too many natural inhibitors, which prevent the PCR from working; if the sample has been improperly stored, or if the DNA has degraded below detectable levels.

Antigen Tests

An antigen test is another way to measure the Lyme disease bacterium. It is most reliable when urine is tested. The test holds great promise as a source of early information because antigens, the foreign substances that stimulate an immune system response, are obviously present before antibodies are created. In addition, antigen may be shed in larger amounts during treatment, as the bacteria die, increasing the potential for positive results. The most accurate results are obtained if a patient takes antibiotics for several days prior to testing, allowing killed bacteria to be detected in sufficient quantities.

The main drawback to this test is that people do not shed antigen regularly, so the test may have to be repeated several times to be accurate. Prolonged antigen shedding may indicate persistent infection.

6

THE TREATMENT OF LYME DISEASE

Antibiotics are the most effective treatment available for Lyme disease, but if you are to be properly treated, you need more than just a prescription for medicine. Good treatment also means forming good relationships with health care providers, developing a strong support network, and having health insurance that pays for the care you need. To get the treatments that may make you well, you have to become an assertive but cooperative patient and an informed, educated consumer. Reading this book is a great first step. Finding other people who have Lyme disease and are willing to share information is also important. I recommend that you subscribe to the Lyme Disease Foundation newsletter, which is published six times a year, to stay current as scientific information changes.

CHOOSING YOUR HEALTH PROFESSIONALS

No single medical specialist has the ideal training to handle a case of Lyme disease, but I recommend that you seek medical care from your family physician, internist, or pediatrician. These experts are trained to take care of the whole body, and they are generally the individuals who know you and your full medical history best. Sometimes, your physician may recommend a consultation with a medical expert more familiar with Lyme disease, such as another internist, a rheumatologist, or an infectious disease specialist.

Because people with lingering Lyme disease spend so much time in medical offices, it is vital to find health care providers you trust. In addition to a physician, you may have to evaluate physical therapists. If you are advised to use intravenous antibiotics, you'll need to find home health care providers as well.

You should also make an effort to find a good pharmacist who is affordable and who accepts your health insurance plan. Establish a relationship with a single pharmacist, and get all your prescriptions filled in one place. Pharmacists are knowledgeable people who can suggest the best ways to take medicine (for example, with meals or on an empty stomach) and will monitor your prescriptions to safeguard against dangerous interactions.

If you are looking for a new doctor, ask some basic questions about office procedures, billing, and medical philosophy before scheduling your first appointment. If a friend is recommending the health professional, find out as much as you can from that person. Then, call the office to request more information. Finally, if you still want to know more, ask for an opportunity to talk to the doctor by telephone before scheduling an appointment.

Here is what you need to know:

What are your fees? What insurance coverage do you accept?

Some doctors charge several hundred dollars for a Lyme disease consultation, and some may expect to be paid at the time of the visit. Others are willing to bill your private medical insurance company and may accept its reimbursement as full payment. Still others will negotiate with you for the difference. Very few private physicians accept Medicaid as full reimbursement but some do, so don't hesitate to ask. Most physicians do accept Medicare, but you may or may not be asked to pay the difference between the Medicare reimbursement and the doctor's bill.

Very few people ask about fees in advance, but you are well within your rights to do so. You should also ask about the costs of any recommended medical tests.

How much do you know about Lyme disease?

New scientific research into Lyme disease and other tick-borne diseases is being published every day, so it is important that your physician know what is happening in the field. Ask whether the doctor attends medically accredited conferences, such as those sponsored by the Lyme Disease Foundation, or subscribes to relevant, peer-reviewed journals, such as the *Journal of Spirochetal and Tick-borne Diseases*. The physician you choose should have a regular attendance at these conferences.

Do you have any potential conflicts of interest I should know about?

You have a right to know if your physician has any financial interest in the firms that provide diagnostic services or treatments for Lyme disease. Some physicians make financial arrangements with diagnostic laboratories or the home health care firms that provide antibiotic infusions. For example, your doctor may own company stock, get paid to review cases for insurance companies, receive bonuses or incentives for making referrals, or be on the company payroll. None of these factors will necessarily influence the recommendations made about your medical care, but you have a right to know about them. If your physician refuses to discuss the matter, you know you are in trouble!

Does my insurer offer you any financial incentives to restrict care?

A doctor-patient relationship should be as private as possible, but in the era of managed care, third parties sometimes get involved in the decision making. I am aware of certain instances when physicians contact a patient's insurance company to find out what treatments are covered and use that information as a guideline. While it is important to consider a patient's pocketbook, it is troubling when an insurance representative without medical training ends up calling the shots.

Some health maintenance organizations (HMOs) forbid physicians to reveal certain information to patients even if it could influence their treatment decisions. In an article in the January 18, 1996, issue of *Time,* one doctor described an HMO's "gag" clause that barred participating physicians from disclosing the full range of testing and treatment restrictions. Many HMOs also offer financial rewards to physicians for keeping treatment costs as low as possible. Several states are passing laws against gag rules, but for now, *caveat emptor.*

How long will it take for me to get an appointment?

Some specialists have a 2- or 3-month wait for a first appointment. You may be able to speed up the process by asking your primary care doctor to intervene and request an earlier appointment.

How long will my appointment last?

Some physicians will spend a full hour with you, taking a complete medical history, reviewing your medical records, and examining you thoroughly. Others spend just 10 minutes with each patient. You can guess who is doing a more thorough job.

How many cases of Lyme disease have you diagnosed and treated?

Doctors who have a lot of Lyme disease experience are more likely to understand the variation of symptoms and the possibility of treatment failure. I recommend seeking out a physician who has treated at least 30–40 cases.

How do you reach a diagnosis of Lyme disease?

Make sure the doctor understands that Lyme disease manifests itself in many ways, depending on the stage of infection, and that the laboratory tests to detect *Bb* antibodies are not necessarily definitive. Ask which blood tests the doctor prefers, and why, and talk about how the results will be interpreted.

Do you have any preconceptions about appropriate treatment?

Perhaps the most controversial question in Lyme disease research centers around why everyone is not cured with 2 to 4 weeks of antibiotic therapy. With all the uncertainties, the only doctor you can trust is one who believes that treatment must be custom-tailored to your needs. Avoid any doctor who claims that everyone except malingerers is cured with 4 weeks of treatment—that is just plain wrong. But stay away as well from medical professionals who believe all patients should have prolonged antibiotic therapy as soon as Lyme disease is diagnosed.

IN THE DOCTOR'S OFFICE

Once you have chosen the health care providers you need, don't second-guess yourself. If you have chosen a physician carefully, don't jump ship just because you are impatient for a full recovery. It is best to establish long-term relationships with professionals who will watch for unexpected changes in your condition.

At the first appointment with a new provider, carry copies of your records, including pictures of any unusual rashes you have had and the results of any laboratory tests. Patient input is critical to getting an accurate diagnosis and appropriate treatment. Keep your appointments, take your medicine as directed, and follow the doctor's orders. Respect the doctor as a valuable health care partner. Missing appointments is disrespectful and suggests you are not committed to improving your own health. If you are cooperative and reliable, your physician is more likely to work hard to help.

Our doctors tend to be pressed for time, so write down your questions before each medical visit. You should never be made to feel uncomfortable about asking anything on your mind. Nor should the doctor ever doubt your word. Unfortunately, people with Lyme disease are frequently dealt with carelessly. If you do not feel that your complaints are being taken seriously, or you feel that inappropriate remarks have been made about your case, it is your responsibility to say so. Tell the doctor or nurse, diplomatically, that their comments are offensive. Ask if they would make similar remarks if you were asking whether a lump might be breast cancer. If the behavior continues, find another physician.

It is very important that you document key developments of your disease. In the appendix of this book I have provided a Personalized Medical Log to help you keep track of key events—such as when you were first bitten by a tick, when you noticed a rash, when you called a doctor, when your symptoms developed, and how the illness has changed over time. Bring your medical log with you every time you see the doctor, and use it to record the dates and locations of any medical tests, such as X rays or magnetic resonance imaging (MRI) scans. You should also make brief notes after every appointment or telephone conversation. Remember to keep copies of any correspondence, including faxes, that you exchange. Ask for copies of important medical records such as test results. This information is invaluable if you switch doctors or see more than one physician.

THE CHALLENGE OF ANTIBIOTIC TREATMENTS

If you have a visible EM rash and begin treatment promptly, up to 4 weeks of oral antibiotics may be enough. The most commonly prescribed oral antibiotics are doxycycline, from the family of drugs known as tetracycline; amoxicillin, part of the penicillin family; and Zithromax, a member of the macrolide family.

Whether or not they receive treatment, some people who have been infected with *Bb* develop signs and symptoms of disseminated disease. We aren't sure if this reflects a wrong antibiotic choice or the fact that the disease had already spread to various organs of the body before treatment began. If illness persists, another round of oral antibiotics is generally prescribed, assuming no neurological problems have become apparent.

If further treatment is required, your physician may recommend that you switch to intravenous antibiotics for 4 to 6 weeks. Unfortunately, even that may be inadequate. Although the best treatment regimen for Lyme disease that has not responded to conventional therapy remains unknown, your doctor can choose from an arsenal of antibiotics and may want to try drugs in different combinations or in different doses. Some concern has been expressed about long-term treatment with oral antibiotics. But in a 1995 study, the National Institutes of Health reported that rheumatoid arthritis patients who used the antibiotic minocyline benefited from significantly reduced joint pain and swelling without experiencing significant toxicity or problems with superinfection.

Your specific signs and symptoms will help determine what medication is recommended. Certain antibiotics distribute themselves more widely in the body, including hard-to-reach areas such as deep tissue, the eyes, and the brain. Some antibiotics are able to go *inside* your cells to attack the bacteria residing there, whereas others are not. Tissue that is inflamed is actually more permeable to certain antibiotics. For example, antibiotics are better able to cross the barrier between the bloodstream and the brain in patients who have meningitis, which is an inflammation of the covering of the brain.

Unfortunately, treatment failures have been reported with every medical regimen currently in use. (For more information about treatment failure, visit some of the Web sites listed in the appendix.) Because we have no tool for determining whether *Bb* has been entirely eliminated from the body, it is hard to be certain what products to use, in what dosage, how they should be administered, and how long treatment should last. The failure to recognize concurrent infections or the presence of other illnesses can add to the challenges.

According to researcher Vera Preac-Mursic in her 1993 book chapter "IV Therapy of Lyme Borreliosis," factors that contribute to treatment success or failure include the degree to which a certain bacterial strain is sensitive to a specific antibiotic, the stage of disease at which treatment starts, the virulence of the strain that causes disease, the amount of bacteria that has been injected into your system, and the places where the bacteria sequester. The fact that strain variations can occur in the same local area further complicates treatment decisions, as do the slow reproduction time of *Bb* and its capacity to remain dormant for long periods, as antibiotics are ineffective against dormant bacteria. Lastly, in 1996 she published her discovery that the Lyme bacterium is able to convert itself to an "L-form" if it encounters a hostile environment. This form of *Bb* has no cell wall and can't be killed by cell-wall activated antibiotics.

In just one study that illustrates the treatment challenges, Preac-Mursic, publishing in the journal *Infection* in 1996, tested various antibiotics on 20 different strains of *Bb* in animals and then tried to culture the spirochete. She found that strains of the bacteria that appeared extremely similar had dramatically different responses to each antibiotic. While Rocephin® worked very well on one strain, a very similar antibiotic, Claforan®, did not. Doxycycline and amoxicillin worked equally well on another strain but azithromycin didn't work at all. Still another *Bb* strain responded to Zithromax better than to Rocephin® or Claforan®.

Results may also differ depending on whether an antibiotic has been tested in the test tube or in animals. For example, erythromycin effectively kills bacteria in the test tube but not in animals, according to an article in the *Journal of Antimicrobial Chemotherapy* published in 1990. In fact, high doses of erythromycin kill the animals being studied before they eliminate all the bacteria. Yet erythromycin is still being prescribed to young children, patients who are allergic to penicillin, and pregnant women. There is no excuse for prescribing an antibiotic that doesn't work, but it is included in some treatment recommendations anyway, giving physicians a false sense of security.

Unfortunately, the recognition that Lyme disease is more difficult to treat than was originally recognized comes at a time when antibiotic-resistant bacteria are beginning to emerge. As a result, there is a strong push within the medical community to reduce the use and length of antibiotic use. In this climate of concern, some physicians may be reluctant to prescribe antibiotics for more than a few weeks even when medically necessary.

How Antibiotics Are Administered

Antibiotics work in two basic ways—either by killing the bacteria directly (bactericidal) or by interfering with the bacteria's ability to reproduce (bacteriostatic), giving your immune system time to kill the bacteria. The limitation of bacteriostatic antibiotics is that once medication stops, bacteria remaining in your body can begin to grow again. Bacteriostatic antibiotics also depend on the productivity of your immune system. However, these two methods of action are not mutually exclusive. Prolonged use of a bacteriostatic drug may directly kill a microbe, even when a bactericidal drug fails.

Antibiotics may be administered orally, through the muscles, or through the veins. If you have a strong desire not to use intravenous antibiotics, you should discuss this subject with your doctor. Also mention any impending personal plans that might affect your treatment options. If you are booked on a tropical vacation, you will probably want to avoid medication that can increase your chance of a sunburn. On the other hand, if your insurance is ending soon, this may be the only opportunity to be reimbursed for costly intravenous therapy.

Oral Antibiotics

While some oral antibiotics are poorly absorbed because the body's stomach acids can inactivate them, others have been designed to avoid that problem.

Additionally, protein in the blood may bind with some of the antibiotic, limiting the level of circulating antibiotic available to fight infection. Fairly common side effects of oral antibiotics are yeast infections or intestinal problems. If you are not improving, your physician may order a "peak and trough" test to find out how much antibiotic is available to fight the bacteria.

Intramuscular Antibiotics

Intramuscular administration makes more drug available to the body than oral medication, while releasing the drugs more slowly, but unfortunately at a lower level, than the intravenous approach. While stomach acids do not inactivate the medicine, intramuscular drugs are still susceptible to protein binding.

Intravenous Antibiotics

Intravenous (IV) antibiotics enter the body directly through the bloodstream, making higher concentrations of drug available. Again, stomach acids do not inactivate the medicine, but protein binding can occur. Most likely, you will receive periodic drips in a doctor's office or a hospital, typically lasting about 1 hour, although pumps are available for 24-hour administration to give you a continuous, level dose of medicine. Many patients learn to administer their own IV at home. If long-term treatment is recommended, a surgical procedure allows a central intravenous line to be placed into a blood vessel leading to the heart.

Intravenous delivery is risky because potentially deadly infections can result. However, if you clean the site at which the IV line enters your body carefully, you are not likely to have a problem. I had an intravenous line in my arm for several months and had no adverse effects.

Choosing an Antibiotic

Antibiotics have a number of potential side effects. Allergies to penicillin and to the closely related category of cephalosporins can be life-threatening. Yeast overgrowths, which may take the form of thrush in the mouth or digestive tract, diarrhea, or vaginal infection, are common side effects of antibiotics, especially if taken over a prolonged period of time. Sometimes, this growth can be curbed by eating yogurt regularly or using acidophilus, a supplement available in many health food stores.

Cost is inevitably a factor in health care decisions. It is possible that an optimal treatment may be beyond your budget or above the reimbursement caps set by your health insurance company. The cost of some oral antibiotics, such as Zithromax, can be several hundred dollars for a several-week treatment, and intravenous antibiotics can cost thousands of dollars a week.

Regardless of what is prescribed, be sure to take the antibiotics for the full length of time your doctor recommends. Even if you begin to feel better, some bacteria may be left in your system and may reproduce as soon as you stop taking medicine. Find out whether the drug is best taken with meals or on an empty stomach. Alcohol may diminish the effectiveness of antibiotics.

Be aware of the Jarisch-Herxheimer reaction, which sometimes occurs as an antibiotic begins to clear the spirochetes from your bloodstream. Accompanying fevers, chills, fatigue, and other aches and pains may actually make you feel worse in the first few days of taking an antibiotic. This condition also occurs with other spirochetal infections and should not be a cause for alarm.

Tetracyclines

Tetracyclines are bacteriostatic, acting by interfering with the bacteria's ability to reproduce, but at higher doses they may actually kill the bacteria. Doxycycline and minocycline, long-acting members of the tetracycline family, are the treatments of choice for many physicians because they are well absorbed on an empty stomach and can be used at lower doses. This class of drugs is also effective against other tick-spread diseases, including relapsing fever, tularemia, ehrlichiosis, and Rocky Mountain spotted fever. Tetracyclines penetrate the cells, making them effective against intracellular bacteria, but levels in the spinal fluid are low, even when brain tissue is inflamed.

Tetracyclines cross the placenta and are not recommended for pregnant or breastfeeding women or for children under the age of eight because they interfere with bone and cartilage development and may discolor or deform developing teeth. Adults should stay out of the sun because this class of antibiotics significantly increases the chance of getting a nasty sunburn in as little as 1 hour. They may be inappropriate for people who spend a great deal of time outdoors, especially in the summer. In addition to yeast infections, other adverse effects can include a mild rash and gastrointestinal disorders, including nausea and vomiting. Rarely, a hypersensitive individual will experience burning of the eyes or liver and kidney disorders. Dairy products, antacids, and iron preparations interfere with the absorption of all tetracyclines and should be avoided.

Doxycycline: As much as 90% to 100% of this antibiotic, which is marketed as Vibramycin® and Doryx®, is absorbed by an empty stomach. It is widely distributed in tissue, with some reaching the spinal fluid and the brain.

Minocycline: Most of this antibiotic is absorbed by the empty stomach and it is widely dispersed into the body. Some antibiotic does reach the spinal fluid and brain. This drug can cause reversible vertigo or dizziness and, rarely, intracranial pressure that can lead to pseudotumor cerebri, the condition that simulates a brain tumor.

Penicillins

This group of antibiotics, originally derived from *Penicillum,* a type of mold, and now chemically synthesized, work by killing bacteria. While the drug will penetrate the lungs, liver, kidney, muscle, bone, and placenta, it does not fight bacteria inside the cells. A very small amount of antibiotic may reach the eyes, brain, spinal fluid, or prostate but *only* if inflammation is present. The degree to which oral penicillin is absorbed differs greatly with each product, but in general, the drug is most effective when it is administered every 4 to 6 hours.

Three members of the penicillin family are commonly used to treat Lyme disease:

Amoxicillin: A drug that is effective against a wide spectrum of pathogens, amoxicillin is well distributed throughout the body. Amoxicillin appears to be fairly safe for pregnant women. A drug called probencid may be prescribed along with amoxicillin to increase the antibiotic level in the bloodstream by delaying its excretion by the kidneys.

Benzathine penicillin G (Bicillin): A single intramuscular shot produces low levels of antibiotics in the blood for between 2 and 4 weeks. It is generally administered once a week in the doctor's office and is rather painful because a large dose of thick medicine must be injected. Family members can be taught to give you the shot.

Penicillin G: Intravenous penicillin G is only effective against the Lyme bacterium if very high doses of the drug are used; unfortunately, only a little drug penetrates the spinal fluid or brain. While some physicians still use this therapy, most have switched to the newer generation of intravenous cephalosporins.

The most common adverse effects of the penicillin family are yeast overgrowths. A mild skin rash, nausea, fever, or joint swelling may also occur, and low levels of the neutrophil white blood cells are sometimes reported. In allergic individuals, penicillin can on rare occasions be fatal.

Cephalosporins

Cephalosporins are derived from *Cephalosporium* fungus and work by killing bacteria. One oral and two IV products have proven to be effective against Lyme disease organisms. The IV products are especially useful for penetrating deep tissue, spinal fluid, and the brain. While a 4-to-6-week course of continuous administration is common, some physicians are trying a several-days-on, several-days-off regimen of IV cephalosporin in an effort to mimic the bacteria's reproduction rate and to avoid adverse effects.

Cefuroxime axetil: This is the only oral cephalosporin proven to be effective against early infection with *Bb*. Marketed as Ceftin®, it is widely distributed throughout the body, including the synovial tissue, joints, and heart. High levels also cross the placenta. Some drug does get into the brain when it is inflamed. In 1996, this drug received the first FDA approval for use in early Lyme disease.

Cefotaxime: This drug, marketed as Claforan®, is administered intravenously, has lower levels of protein binding than ceftriaxone, and is not associated with gallstones. The drawback to cefotaxime is that it is shorter lasting and must be administered two to three times a day.

Ceftriaxone: Ceftriaxone, marketed as Rocephin®, needs to be administered only once a day. Ceftriaxone may foster gallstones or sludging, but this is generally reversible when antibiotic use is discontinued.

Other possible adverse affects of the cephalosporin antibiotics include diarrhea, rash, and yeast overgrowth. Some of these problems can be life-threatening. A small number of people who are allergic to penicillin are also allergic to cephalosporins.

Macrolides

This group of antibiotics, discovered in the 1950s, incorporates a macrolide ring in its structure. Macrolides work against a broad range of bacteria, preventing them from reproducing and at high doses killing the pathogen. These drugs penetrate the tissue and go inside the cells to battle intracellular infections. Although less effective for penetrating the brain and spinal fluid, they are considered useful for eye infections. Macrolides cross the placenta and are excreted in breast milk. Antibiotic levels in the tissue are 10 to 100 times higher than in the bloodstream.

Azithromycin: This oral antibiotic, marketed as Zithromax® has the highest tissue penetration of the macrolide group and lasts for several days in tissue. Food in the stomach reduces absorption by as much as 50% so this should be taken on an empty stomach. Plaquenil® (hydroxychloroquine), an anti-inflammatory drug, is often prescribed in conjunction with Zithromax to treat other diseases. Dr. Sam Donta, chief of infectious diseases at Boston University, uses Plaquenil to change the intracellular pH (less acid) to allow the macrolide to be more effective. This combination therapy is referred to as Z.A.P. (Zithromax and Plaquenil). Azithromycin is rapidly cleared from the bloodstream.

Clarithromycin: A recent study of clarithromycin, marketed as Biaxin®, was published in *Antimicrobial Agents and Chemotherapy* and suggests that this oral antibiotic can work very well against Lyme disease. One possible reason is that the antibiotic is widely distributed, and levels remain high both within cells and in the bloodstream.

Potential side effects of the macrolide family of antibiotics includes gastrointestinal problems such as cramps, nausea, vomiting, and diarrhea, and skin rashes. Occasionally, a reversible hearing loss may occur. Rarely, there is liver damage or an increase in white blood cells.

When Should You Use Antibiotics?

Because of the anxiety prompted by Lyme disease and other tick-borne disorders, some people request antibiotics before venturing into tick-infested areas, especially if they are camping, hiking, or hunting. There is some precedent for this—for example, people with certain heart conditions use antibiotics prior to visiting the dentist to safeguard against infection. Antibiotics may also be used as to prevent traveler's diarrhea.

Still, prophylactic antibiotics are controversial, and I do not recommend them. Chances are they will be unnecessary, and you could be contributing to the problem of antibiotic-resistant bacteria, which is occurring in part because these drugs have been overprescribed.

I do recommend empiric therapy, where you begin treatment as soon as you discover that you have been bitten by a tick and may have been exposed to a pathogen, especially if you are in an area where infection is endemic. Not everyone agrees with this approach, either—some people argue that no medicine should be used unless there is an active infection. This "watch and wait" approach was routine until new scientific studies and emerging public health concerns began to challenge this convention. Today, we realize the delaying treatment may result in an infection that is significantly harder to treat.

In light of the risks, I think a "better safe than sorry" approach is prudent. On the one hand, you may be taking twenty dollars worth of medications unnecessarily if the tick was not infected. On the other hand, you may be avoiding a nightmarish scenario that involves thirty thousand dollars worth of intravenous medicine. My recommendation matches the scientific wisdom expressed in *Clinical Use of Antimicrobial Therapy,* in which author Steven Barriere, PharmD., advises that empiric therapy, in which infection following exposure to a disease-causing agent is presumed, is appropriate when an untreated infection carries a significant risk of disability before a laboratory can identify the pathogen involved. Empiric therapy is initiated for many diseases. Given the early dissemination of the Lyme bacterium throughout the body and into the brain, I think that waiting for a telltale EM rash is just bad medicine.

Other reasons for treating when you discover that you have been bitten by a tick are that you are unlikely to know what type of tick has bitten you, how long it has been attached, and whether it is infected with one or more pathogens. Early treatment is especially advisable if you have been bitten in an infested area, although many communities have not fully assessed this danger or are sitting back in blissful ignorance. People in high-risk groups, including pregnant women, babies, young children, and anyone with a serious illness should definitely consider early treatment.

Early treatment may also be the most economical approach. In "Prevention of Lyme Disease After Tick Bites: A Cost-Effectiveness Analysis," published in 1992 in *The New England Journal of Medicine,* the authors argued that in an area where 30% of the ticks were infected with *Bb,* providing treatment when a tick bite occurs is the least expensive strategy and results in the fewest cases of Lyme disease with the fewest complications.

My advice is to discuss the subject with your physician so that you can make an educated decision. An antibiotic administered when a tick bite occurs is given at the same dose, and for the same length of time, as it would have been had the EM rash occurred. The first line of defense will generally be doxycycline.

INSURANCE COVERAGE

With managed care and HMOs playing an increasingly dominant role in the nation's health care system, many of us are finding that our insurance companies have almost as much to say about the treatment we receive as our own doctors do. Your health care plan may require you to see one or more physicians on multiple occasions and to undergo a number of laboratory tests. In order to get the level of care you need, you may have to negotiate, or argue, with your managed care organization. Coverage for prescription drugs may also be denied, especially if a recommended medication has not been specifically labeled for the treatment of Lyme disease. Many insurance companies consider this "off-label use" to be synonymous with "experimental" use, even if a prescribed medication has actually become the standard of care preferred by most physicians.

It is very important to familiarize yourself with the terms of your own policy. Read the fine print to learn more about your coverage and how much you will be reimbursed for your care. Be sure to find out if you need to receive prior approval (called "pre-certification") for certain treatments or tests.

If a plan administrator tells you that a certain procedure is not covered or an HMO refuses to allow you to consult a specialist, be prepared to assert yourself. Most health care plans have more flexibility than patients realize. For example, if you can demonstrate that your insurer reimburses for other off-label drug uses, you may strengthen your case for arguing that your own medicine should be covered.

Resolving differences of opinion can be costly and time-consuming, but the battle is worthwhile if you can persuade your insurer to foot the bill for your medical coverage. Here is what you need to know to win.

When Coverage Is Denied: First Steps

If your health insurance company refuses to pay for the treatment you want or need, you have a number of tactics at your disposal. First, be sure that your complaint is legitimate. Sometimes, health services *are* overused and your insurance company may be prudent in refusing to cover unnecessary treatment. However, if your physician recommends appropriate medical care that your insurance company refuses to pay for, your grievance is probably reasonable.

Next, give the insurer a chance to fulfill its commitment. Sometimes, a problem turns out to be a matter of miscommunication or incomplete documentation. Make sure your communication with your insurer is clear.

If coverage is denied, get it in writing. Then, contact your employee benefits department, if you have one, and ask for their assistance. Most employers contract with one or more insurance companies to provide coverage, but they seldom relinquish the full reins of authority. And most insurance companies want to satisfy your employer, knowing that other health coverage options are available if complaints begin to mount.

Send a registered letter to the insurer, return receipt requested, with a copy to the employee benefits department, requesting a reevaluation of your case and clarifying the medical necessity for the sought-after care. State that the insurer will be held liable for any hardship or injury incurred due to the delay in diagnosis or treatment. Include a copy of a letter from your doctor explaining the medical reasons for the diagnostic tests or treatments that are being denied. You may also want letters of support from other physicians confirming that the treatment recommended by your doctor is appropriate. A copy of the insurance company's denial of coverage should also be enclosed. Ask for a response within 5 days.

Contacting the Insurance Commission

If you still don't get satisfaction, contact your state Insurance Commission (in some states, it is called the Insurance Department). You can find the commission by logging on to the Lyme Disease Foundation's Web page (see appendix), by checking the government pages of your local telephone book, or by asking Directory Assistance. As the regulatory body with authority over private companies, the Insurance Commission can help you reestablish the lines of communication with your insurer, move a dispute into mediation, or ensure that the terms of your health insurance policy are respected. When you complain to the Insurance Commission, the board of directors at your own insurance company receives a copy of your letter and is required by law to respond.

The state Insurance Commission has no authority over the health plans of large corporations that are self-insured, which are discussed below, or over disputes with Medicare. Also, it cannot force an insurer to pay expenses that are not part of the health care contract and it must step out of the picture if you file a lawsuit.

The package that goes to the Insurance Commission should include the following:

Cover Letter

A cover letter should include this background information:

- the policy number, plan number, and ID number of the relevant insurance plan
- the insurance company's name, address, and telephone and fax numbers
- the insurance plan administrator's name, address, and telephone and fax numbers, if different
- the insured's name, address, and telephone and fax numbers (this could be you, your spouse, or another family member)
- the name of the patient involved in the coverage dispute
- your relationship to the insured and to the patient, if different
- your name, address, and home and work telephone and fax numbers
- the best time and place to reach you

Case Summary

In no more than one or two pages, concisely state your problem and the action you want taken. Ask the Insurance Commission to use all its resources to ensure that the insurer meets its legal obligation to provide coverage for necessary medical care. Request the names and qualifications of the insurance company's "expert" consultants, the fees paid for their opinion (including the expert who established the company's overall guidelines), and the criteria used to decline your coverage. You should also ask that the insurance company promise, in writing, to comply with the terms of your policy, and not to harass you or your physician.

Supporting Evidence

Include any correspondence or telephone logs from conversations with your physician that support the medical necessity of the disputed care. A copy of the letter from your insurance company declining coverage or questioning the necessity of treatment should also be included, with relevant passages highlighted.

Stepping Up the Pressure

When you file a complaint with the Insurance Commission, send copies to your congressional representatives and ask that they assist you in obtaining full insurance coverage.

You might also want to go public by sending the package to the local and national media. Corporations pay a fortune for good public relations and are fearful of bad publicity. If a reporter makes a telephone call inquiring about your case, you may see some fast action.

If all else fails, you might consider filing a lawsuit against your insurance company. Sometimes, the mere threat of legal action, in the form of a single letter from an attorney, is enough to resolve a conflict. Lymenet on the World Wide Web (see appendix) includes a legal section of Lyme disease–related court decisions that may help you.

Employers' Self-Insured Plans

Some large corporations insure themselves by putting their own money into a special fund and paying health claims from this fund; this is called "self-insured." A professional claims manager, or a company contracted for the purpose, generally administers the plan and makes decisions about coverage. Such plans are regulated under the federal Employee Retirement Income Security Act of 1974.

If you disagree with a decision, try talking directly to the claims manager, and ask what you should do to obtain the care that you need. If you don't make any progress, you may get assistance from the employee benefits department. Remember, it is in your employer's interest to keep you healthy, and most companies retain at least some control about how a self-insured plan is administered.

If you still can't work out your differences, find out how to appeal. The first step is generally to ask for the "denial of coverage" in writing. It is also essential to document each of your conversations with your health care providers and claims manager. Your rights and responsibilities during an appeal are probably spelled out in the company booklet that outlines the terms of your coverage.

If you need further assistance, contact the Pension and Welfare Benefits Administration at the local office of the U.S. Department of Labor.

Obtaining Your Shared Insurance File

The Medical Information Bureau (MIB) was organized in 1902 by physicians and insurance companies, originally to reduce fraud that led to higher health care premiums. Its main purpose now is to prevent consumers from omitting or concealing medical information when they apply for medical insurance. Information is obtained from numerous health care sources, including providers and clerks working in private offices, hospitals, laboratories, and other medical facilities. Today, virtually every insurance company belongs to the MIB, and if you have applied for health or life insurance in the last seven years, chances are that you have an MIB record to which insurers are allowed access. When you apply for new insurance, the information you provide is compared with information already in the MIB's medical files.

Until recently, consumer access to this information was limited, but now you can request a copy of your MIB medical record. Write for an application from the bureau (see appendix for the address). Once you complete and submit the required forms, and pay the eight-dollar fee, which is mandatory even if no record is found, you will receive a complete copy of your file within 30 days. This record, which has all MIB codes translated, allows you to see what information has been collected about you, which insurance companies have received copies of your files, and when. If you find and report mistakes, MIB will correct them.

The bureau receives about fifty thousand requests for information annually, and about eight hundred people ask for corrections. If you have applied for insurance in the last 7 years, I strongly recommend that you write for a copy of your files and correct them promptly if you find any errors.

MAXIMIZING YOUR SUPPORT SYSTEM

Taking care of your body is only part of securing the care you need. You may also need emotional support to deal with the personal and professional upheavals associated with chronic illness. Your physician can't help you cope with these stressors, and even family members are likely to have their limitations, no matter how much they love you. Most likely, they cannot understand exactly what you are going through and may be dealing with their own fears as a result of your illness.

I recommend a good Lyme disease self-help group for anyone suffering from a tick-borne disorder. Self-help groups carry the message of recovery and hope to people who are suffering. They provide positive support when your life may seem to be falling apart. It is not a sign of weakness to join and attend meetings. Quite the contrary—joining a self-help group is a positive sign that you recognize your own needs and value yourself enough to try to meet them.

Nonprofessional, volunteer-led support groups offer strategies for coping with serious problems and sometimes a sympathetic shoulder to cry on. Most important, these groups help eliminate your sense of isolation and remind you that you are not alone. Through the advice and support of others facing the same challenges, you can empower yourself and regain control over your life. Family and friends are usually welcome to attend and may as a result gain valuable insight into your ordeal.

In-Person Self-Help Groups

A typical support group consists of a *facilitator,* who runs the meeting and guides the discussion, and a *coordinator*, who arranges the advance details, such as the meeting time and location and preparation of any materials to be distributed. Most groups are informal and are run entirely by committed volunteers.

After a general introduction and hello, the facilitator usually reviews the group's mission for the sake of any newcomers and introduces the other volunteers who are helping out. Depending on the size of the group, visitors may then be asked to introduce themselves and to describe their experiences with Lyme disease. A guest speaker is often invited to make a presentation. Then the floor will be thrown open so that everyone present has the opportunity to ask questions, add their own insights, and describe their own struggles. Respect for confidentiality is a very important principle of these meetings, and members are allowed to remain anonymous if they wish.

As the meeting closes, the time and place of the next meeting are generally announced. Most meetings end on an upbeat note. You should feel better and more hopeful as you leave. If you do not, make it a point to talk about your experiences with the facilitator.

Cyberspace Self-Help Groups

Computer chat groups are especially valuable for people who are homebound or who live far away from an in-person support group. Both CompuServe and America Online have well-attended self-help chat rooms for people with Lyme disease and other tick-borne disorders. Guests, such as a doctor or researcher, are sometimes invited, but more often, there are free-floating discussions, with lots of people posting and responding to messages at the same time. Cross-conversations are commonplace. (See the appendix for a complete list of computer chat groups.)

Professional Counseling

The expert help of a professional counselor is sometimes appropriate, especially if you have emotional difficulties that you do not want to share with the people in your support network. If you are angry, depressed, or highly stressed over a prolonged period of time, think seriously about consulting a psychologist, a social worker, or a psychiatrist. Psychiatrists are medical doctors who can prescribe medication to help you cope. Other professional counselors focus more on getting you to think about, and talk through, your problems so that you can get back in control of your life.

People who must deal with chronic illnesses sometimes have suicidal thoughts. If you ever feel as though ending your life is the only way to reduce your emotional or physical pain, you must seek immediate help. This is a very difficult issue to deal with, but you cannot afford to delay.

Caring for Yourself

In addition to seeking conventional medical care and building a strong support system, patients with Lyme disease can do a great deal for themselves:

Practice Good Nutritional Habits

Give your body all the help you can by eating healthy foods and cutting down on high-fat and fast foods. Eliminating alcohol and caffeine from your diet may also boost your energy. You may want to seek the advice of a professional nutritionist.

Use Alternative Therapies with Care

Some patients have reported success with unconventional healing methods, but I urge you to talk with your physician before experimenting. Vitamins taken at reasonable doses are unlikely to do you any harm, and may be helpful, but some other remedies are much less benign. Some patients tried malaria therapy, in which they deliberately exposed themselves to malaria, hoping to raise the body temperature high enough to eliminate Lyme disease bacteria. Unsurprisingly, this technique didn't work, and several patients almost lost their lives in the process. If you choose to try alternative therapies, be sure to continue your antibiotic regimen at the same time.

Modify Your Lifestyle

It is very important to decrease the stressors in your life and set aside time to rest. Give yourself an opportunity to make a full recovery, even if that means postponing chores and modifying your work schedule. You may even have to change jobs, at least temporarily. If your job requires physical exertion, for example, it may be time to ask for a temporary reassignment to a desk job.

Ask Your Family to Help

Don't hesitate to ask family members to pitch in. Some of us are reluctant to ask for help, but we are entitled to make demands of the people we love, as long as they are fair and are openly discussed. If you plan carefully and make sure other members of your family have input when routines change, you can minimize the disruptions and give everyone the sense of being partners in your recovery.

Keep Your Sense of Humor

Laughter really is the best medicine. In the absence of definitive answers, life can sometimes seem scary and hopeless, but behind every cloud there really is a silver lining. Take the story told to me by Edward Masters, a Missouri physician. His patient suffered from short-term memory problems linked to neurologic Lyme disease, which she sometimes found useful. Why? Because it allowed her to hide her own Easter eggs!

7

OTHER TICK-BORNE DISORDERS

Unsettling though it is, the same minuscule tick that spreads *Bb* can also transmit other illness-causing pathogens through the same bite. Some patients with Lyme disease also have babesiosis, ehrlichiosis, or all three. As you might imagine, patients with multiple infections typically have more severe symptoms than those with a single disease. Other tick-borne diseases, including Rocky Mountain spotted fever, Colorado tick fever, tularemia, relapsing fever, Powassan encephalitis, and tick paralysis can also cause severe illness and occasionally death.

Some of these illnesses are reasonably well understood. Others have only recently been identified, and some important information about routes of transmission, diagnosis, treatment, or long-term course of illness is still missing. What is included here is, of course, as up-to-date as possible, but science is advancing rapidly, so you should be on the alert for new discoveries. I recommend that you clip articles as you find them and file them in the back of this book in order to keep your reference material current and well organized. The brief history in the disease descriptions that follow is intended to put them in context and to help you understand how scientific knowledge evolves.

BABESIOSIS

Babesiosis is caused by *Babesia microti* and other species of *Babesia,* protozoa that invade, infect, and kill the red blood cells. In North America, human infection has mostly been reported in coastal and island areas and in the upper Midwest of the United States. The disease has also been reported in Europe.

In 1957, a Yugoslavian researcher named Skrabalo was first to describe babesiosis as a human disease. The first reports of infection in North America came 12 years later, on Nantucket Island in Massachusetts. About 100 cases were voluntarily reported through the mid-1990s, but public health officials and physicians fear that the incidence of disease may be far more widespread than was initially suspected. This fear is strongly supported by a 1996 study in Connecticut that showed that as many as half of all *Bb*-carrying ticks were also infected with *Babesia*. Especially alarming was the 1996 discovery of a deadly strain of babesiosis in Missouri, currently known as MO1, which appears to be closely related to dangerous European strains. Other *Babesia* species have recently been identified and described, but they are not yet named, and little is known about them.

How Is Babesiosis Transmitted?

Babesia is transmitted by the black-legged tick, which acquires the infection from feeding on an infected host, most often small, white-footed mice or other rodents. These ticks stay infected throughout their lives but cannot pass infection to their eggs. The Western black-legged tick is under study as a possible vector. Blood transfusions can also spread the infection.

What Are the Symptoms?

Symptoms typically become apparent within 1 week of a tick bite but may take as long as 6 weeks to manifest themselves. Disease usually begins gradually, with flu-like symptoms that linger, sometimes for weeks. The symptoms are generally mild at first and may be accompanied by periodic fever. In more severe cases, a patient may experience drenching, malaria-like chills as well as fatigue, headache, weakness, and muscle aches or pain. Other symptoms include vomiting, depression, anemia, small amounts of blood in the urine, and low blood pressure. Jaundice, stemming from problems in the liver, is usually recognized by the yellowish cast it gives the skin and eyes. The kidneys or spleen can also be affected by babesiosis.

Babesiosis is generally more severe, and can be fatal, in elderly people or those who have had their spleens removed.

What Diagnostic Tests Are Available?

Patients with babesiosis are sometimes misdiagnosed as having only Lyme disease, but a number of diagnostic tests can reveal the truth. A standard blood test often shows a breakdown of red blood cells, a decrease in platelets, which are responsible for the blood's ability to clot, and elevated liver enzymes. In a more targeted test, a lab technician can stain a smear of your blood and then look for the characteristic protozoa through a microscope. The stained protozoa will appear as ring-shaped objects inside the red blood cells, but this test is not foolproof—even if the technician does not find the protozoa, babesiosis can not be definitively ruled out.

In another diagnostic test, the level of antibodies being produced to combat the *Babesia* protozoa is measured at the time of acute symptoms and then compared to antibody levels during recovery. If there is a fourfold difference, a patient can be definitively diagnosed with babesiosis.

A polymerase chain reaction (PCR) test for DNA is also available.

What Is the Treatment?

Clindamycin and oral quinine are generally used to treat babesiosis. In cases of very severe infection, your physician may consider a blood transfusion.

EHRLICHIOSIS

Ehrlichiosis is an infection caused by a type of round rickettsiae (a bacterial parasite) that invades and infects the white blood cells, which ordinarily act as scavengers, ingesting and killing foreign invaders. Rickettsiae live and reproduce inside the cells and then kill the cell as they exit. Two severe and sometimes fatal illnesses—human monocytic ehrlichiosis (HME) and human granulocytic ehrlichiosis (HGE)—are caused by a rickettsial parasite in humans. HME targets primarily monocytes and granulocytes occasionally; both are a type of white blood cell. HGE invades and infects neutrophils, a type of granulocyte.

The disease-causing parasite was first discovered in Algerian dogs in 1935 and was soon found on other continents as well. In 1945, when researchers realized they were studying a distinct subtype of *Rickettsia,* the parasite earned the name *Ehrlichia,* after the famous German microbiologist Paul Ehrlich. Initially, this parasite was thought to cause infection only in animals, including dogs, cows, and sheep, but mid-century researchers isolated *Ehrlichia sennetsu* as the cause of an infectious mononucleosis-like illness that had been described in Japanese literature since the late 1800s. The disease was named "Sennetsu fever" and is believed to be transmitted by eating raw fish.

In the Western Hemisphere, the first human case of ehrlichiosis was described in Arkansas in 1986 after grapelike clusters of parasites were found in the monocytes of a white blood cell sample. In 1990, a Centers for Disease Control and Prevention (CDC) researcher named Jacqueline Dawson isolated *Ehrlichia chaffeensis,* now recognized as the cause of human monocytic ehrlichiosis. *Ehrlichia chaffeensis* DNA closely resembles that of *Ehrlichia canis* and *Ehrlichia ewingii,* which infect the white blood cells of dogs.

Over the next few years, physicians in the Midwest noticed that some patients had symptoms that closely resembled those caused by *Ehrlichia chaffeensis* but were not manufacturing antibodies to that parasite. After further investigation, researchers realized that in this population, *ehrlichiae* were clustered in the neutrophils, not in the monocytes, which meant a different tick-borne infection was involved. The new disease was named human granulocytic ehrlichiosis and was found to be caused by *Ehrlichia equi,* which had previously been known only as an agent of disease in horses and some other animals.

Most reports of HME come from the south-central and southeastern United States, but some four hundred cases have been confirmed in more than thirty U.S. states, and the actual incidence of infection is almost certainly far higher. HME has also been found in Africa and Europe. Scientists originally thought that HGE was limited to Minnesota and Wisconsin, but other researchers have shown that this disease is widespread in nature and has infected people in Arkansas, California, Connecticut, Florida, Georgia, Maryland, New York, Pennsylvania, Rhode Island, and Tennessee, as well as in the Midwest. HGE has also been reported in Europe. The full geographic spread of the infection is not yet certain.

At the Connecticut Agricultural Experiment Station, researchers are investigating still another strain of ehrlichia that they believe may cause disease in humans.

How Is Ehrlichiosis Transmitted?

Based on laboratory experiments, ticks are believed to be able to transmit HME and HGE infection after attaching and feeding for 1 to 2 days.

HME: The lone star tick is the primary vector of *Ehrlichia chaffeensis,* which can also be transmitted by the American dog tick and the brown dog tick. The Western black-legged tick may also be able to transmit infection. Common reservoirs of *Ehrlichia chaffeensis* include medium-sized mammals and white-tailed deer. The parasites can actively circulate for as long as 2 weeks in the bloodstream of deer, creating a substantial window of opportunity for feeding ticks to acquire the parasitic infection. Once infected, these ticks remain so throughout their lives, but the adult female probably does not pass *Ehrlichia* infection to its eggs. HME infection is most likely to occur between March and October.

HGE: After feeding on an infected host, the black-legged tick and possibly the lone star tick, the Western black-legged tick, and the American dog tick are able to transmit *Ehrlichia equi* to humans. The ticks remain infected throughout their lives but cannot pass infection on to their eggs. HGE is most likely to be reported in May or June and in the late fall.

What Are the Symptoms?

Symptoms typically begin between 7 and 14 days after a tick bite but may start almost immediately or may not become apparent for as long as 3 weeks. Anemia may be a tip-off of this disease.

HME: While some patients suffer no symptoms of HME, others become severely ill. HME kills between 3% and 5% of those infected, almost always due to an underlying illness that has compromised the immune system or because treatment is delayed or inappropriate. Sometimes infection persists, causing bouts of acute illness interspersed with periods of no symptoms.

Many of the symptoms of HME resemble those of Rocky Mountain spotted fever (RMSF) and may include fever, malaise, headache, chills, sweating, severe muscle aches and pain, and nonproductive cough. Gastrointestinal problems including abdominal pain, nausea, vomiting, and bleeding sometimes occur. There are also reports of anemia and abnormal decreases in white and red blood cells and in blood platelets. Infection can damage most other organs of the body as well. For example, liver enzymes may be elevated, heart rate may slow, pneumonia may be diagnosed in the lungs, and the kidneys may fail. Problems may affect the bone marrow, the spleen, and lymph nodes. Inflammation of the brain, a condition known as encephalitis, has been reported.

About 20% of the time, a rash similar to the RMSF rash occurs. Typically, the HME rash is more diffuse and may include numerous small and discolored spots or patches that are reddish or purple and may look like bleeding underneath the skin. The patches are level, not raised or indented, and can be contrasted with the RMSF rash because they do not appear on the palms and soles of the feet.

HGE: HGE can become severely debilitating within a matter of hours. Multiple body systems may be involved, and about half of those infected must be hospitalized for a week or more. The infection may be deadly to as many as 10% of patients, usually because of improper diagnosis and delayed or inappropriate treatment. Individuals who have compromised immune systems are at special risk.

While the full extent of symptoms associated with HGE is still under investigation, the disease typically begins as an acute onset of severe flu-like symptoms, including high fever, chills, sweating, malaise, severe headache, muscle and joint pain, and a combination of gastrointestinal symptoms, including nausea, vomiting, abdominal pain, or diarrhea. There may also be problems in the lungs, including pneumonia and respiratory failure, as well as meningitis, confusion, and other central nervous system symptoms. Anemia and bleeding complications have been reported, and most patients have mild to moderate liver damage. Kidney damage resulting in increased nitrogen in the blood is common and may lead to total kidney failure, although sometimes kidney function remains normal.

Rarely, the rash described for HME may also occur with HGE.

What Diagnostic Tests Are Available?

The *Ehrlichia* parasites that cause HME and HGE are sufficiently distinct and do not usually cross-react with each other on blood tests. Patients with HGE may test negative to HME tests and vice versa, which means that anyone suspected of having ehrlichiosis should be tested for both parasitic species.

HME: To diagnose human monocytic ehrlichiosis, a laboratory technician will place a blood sample on a small glass slide, stain the slide, and look for the presence of stained *Ehrlichia chaffeensis.* If clusters of small dots (morulae) can be seen inside the cell, the test is considered positive. The most accurate stains are taken during the acute fever stage.

Other diagnostic options include the indirect fluorescent antibody test and the Western blot test. Some laboratories also offer the PCR test. These tests are all described in chapter 5.

HGE: Blood tests to determine whether you are infected with the pathogens that cause HGE may reveal anemia or a decrease in blood platelets, white blood cells, granulocytes, or neutrophils. However, one study of twelve infected patients showed completely normal blood work. To confirm infection, stained *Ehrlichia equi* must be found. Alternatively, a diagnosis of HGE can be made if you have a positive PCR test or have antibodies to *Ehrlichia equi*.

What Is the Treatment?

Both HME and HGE are best treated with tetracycline or doxycycline, antibiotics capable of entering human cells to prevent the bacteria from replicating, which ultimately leads to its death. Chloramphenicol is sometimes used in treatment if there is a reason not to use tetracycline. However, leading researchers warn fellow physicians that chloramphenicol *does not* kill the pathogen. According to the Department of Health in New York State's Westchester County, rifampin may be the best treatment alternative for children.

A full recovery can be expected of patients who are promptly diagnosed and appropriately treated, but fatalities do occur in those who are not. There have also been reports of persistent infection after treatment that is generally considered adequate.

ROCKY MOUNTAIN SPOTTED FEVER

Rocky Mountain spotted fever (RMSP), also called tick-borne typhus, is caused by the *Rickettsia rickettsii* parasite, which invades the cells lining the heart and blood vessels. It is the most virulent of a number of spotted fever diseases and occurs throughout the Western Hemisphere. Physicians are required to inform the CDC when they see patients with RSMF. About five hundred cases are reported annually in the United States.

RMSF was first identified in 1873 in Idaho. By 1906, a Montana researcher named Dr. Howard Taylor Ricketts proved that the disease was infectious and that it was spread by ticks. Polish scientists developed the first blood tests to identify RSMS in 1915.

How Is RMSF Transmitted?

The pathogen for RMSF can be transmitted by the American dog tick, the Rocky Mountain wood tick, and the Pacific Coast tick. Infection is maintained as a tick molts, allowing *Rickettsia rickettsii* parasites to be transmitted at any stage of a tick's life. Adult female ticks can also pass infection to their eggs.

Within a few hours of attaching and beginning to feed, the tick can pass the *Rickettsia rickettsii* parasite into the human body. Because the infection is transmitted in the tick's saliva, it is also possible to become infected if your hands or body come in contact with the tick's mouthparts, especially if you try to crush the tick between your fingers. Tick fluids can also penetrate the mucous membrane or an open sore.

What Are the Symptoms?

Symptoms of RMSF typically develop about 1 week after a tick bite, although they may become apparent within 2 days or as long as 2 weeks later. While some infected individuals experience very mild symptoms, others may be left with serious permanent damage. Disease onset is sudden, generally with a classic triad of symptoms, including a high fever (102°F), severe headache, especially behind the eyes, and a hallmark skin rash, which begins with numerous small spots on the wrists, palms, ankles, and soles of the feet. The spots are the size and shape of measles (see Figure 7–1) and on light skin they are typically reddish to black or look like blood under the skin. Gradually,

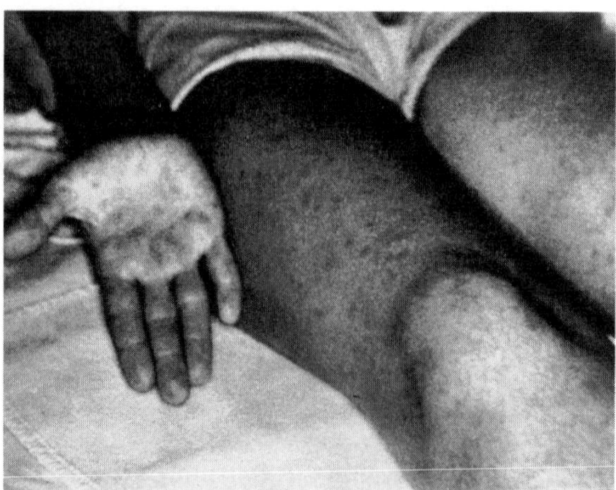

Figure 7–1. Rocky Mountain spotted fever rash. Reprinted by permission of Texas Department of Health.

these spots spread to the rest of the body, usually sparing the face. In up to half of the patients, the rash is totally absent.

There is significant variation in other reported symptoms. Flu-like complaints may include chills, confusion, light sensitivity, nausea/abdominal pain, and achy or painful muscles. Gastrointestinal problems, including nausea, vomiting, diarrhea, and abdominal pain and tenderness can occur, as can transient deafness and jaundice. Other complications that may arise include lung inflammation or pneumonia, inflammation of the heart, an enlarged liver or spleen, kidney failure, and gangrene, especially on the fingers and toes. Occasionally, circulatory collapse and central nervous system problems, including delirium, occur, and patients may lapse into a coma.

The death rate for RMSF has been estimated at between 3% and 8%, but it is almost always limited to people who have not been treated early or appropriately with antibiotics. Because dark skin may mask the telltale rash, people of color may not notice early signs of infection, which contributes to a higher fatality rate.

What Diagnostic Tests Are Available?

Ordinarily, a physician will make a clinical diagnosis of RMSF on the basis of symptoms and begin treatment immediately. Antibodies to the disease do not appear for 10 to 14 days after the first symptoms, limiting the early value of the blood test. To confirm the diagnosis, however, blood tests can be run during the acute phase of the disease and again as you recover. A fourfold difference in antibody levels is considered proof of RMSF. Ask if the laboratory uses a test specific for RMSF; some labs are using the obsolete Weil-Felix test, which is not as good.

A standard blood count may show lower platelet levels. Some laboratories may try to stain a tissue sample in order to look directly for the *Rickettsia rickettsii* parasite.

What Is the Treatment?

RMSF must be promptly treated with antibiotics, such as tetracycline, doxycycline, or chloramphenicol. An intravenous solution may also be used for patients who have become dehydrated. An infusion of platelets or clotting factor may help resolve blood-clotting difficulties, a respirator may be necessary to assist with breathing, and medication may sometimes be prescribed to help regulate the heart.

COLORADO TICK FEVER

Colorado tick fever is a double-stranded ribonucleic acid (RNA) viral disease that infects red blood cells. Because the virus, a member of a group known as reoviruses, lodges itself inside the cells, antibodies that circulate through the bloodstream cannot reach it. Although Colorado tick fever was first described in 1850, it was initially lumped with other diseases of unknown causes. In the 1930s, it was recognized as a discrete condition. A decade later, once its infectious nature had been recognized, L. Florio was able to isolate the reovirus from Rocky Mountain wood ticks.

Colorado tick fever is prevalent in mountainous regions of the United States, with most cases being reported in Colorado and Utah. Between two hundred and four hundred cases are voluntarily reported each year in the United States, but the actual incidence of disease is widely believed to be much higher. Even in Colorado, most physicians do not realize that Colorado tick fever is a reportable disease, according to officials in the state Department of Public Health and Environment. Available statistics suggest that twice as many men contract the disease as women, probably reflecting the fact that men are more likely to do physical labor in the woods.

In 1991, a variation of the Colorado tick fever virus was discovered in two patients living near the Salmon River in Idaho. Scientific interest was piqued because these people had characteristic symptoms of Colorado tick fever, but the antibody tests were negative, and the virus could not be isolated from their blood. At about the same time, other physicians were seeing patients with symptoms resembling RMSF in Oklahoma and Texas. After probing further, scientists identified two viruses that resembled, but were not identical, to those known to cause classic Colorado tick fever.

How Is Colorado Tick Fever Transmitted?

In the United States, the virus that causes Colorado tick fever is transmitted by the Rocky Mountain wood tick. The Pacific Coast tick and others may also be infected with the virus but have not yet been tested to determine whether they can transmit it to humans. Ticks that are infected remain so throughout their lives, but the female does not pass infection on to its eggs. Small mammals, such as mice, squirrels, deer, chipmunks, and porcupines, apparently maintain the infection in nature. Human infection is most likely to occur from March to November, with the majority of cases reported between April and June. Because blood transfusions may transmit disease, anyone who has been diagnosed with Colorado tick fever is asked not to donate blood for at least 6 months.

Ixodes ticks are known to transmit a variation of Colorado tick fever in France and West Germany.

What Are the Symptoms?

Symptoms of Colorado tick fever generally appear within 3 to 6 days after a tick bite, although some people report symptoms when the bite occurs and others take up to 2 weeks to feel ill. The virus has been found in the respiratory and digestive tracts of some healthy individuals without symptoms.

Colorado tick fever generally begins with a sudden onset of high fever (about 104°F). In about half the cases, the high fever is followed by a remission that lasts 2 or 3 days, and then a second phase of high fever lingers for 2 to 4 days more. The dual-phase fever is a hallmark of this disease. Other symptoms include chills, severe muscle and back pains, headache, especially behind the eyes, and light sensitivity. About 10% of patients also develop a faint and rather nondescript rash. More serious complications include bleeding, heart problems, pneumonia, hepatitis, and encephalitis.

Congenital infection has been documented, but there is no clear association either with spontaneous abortion or with fetal anomalies.

What Diagnostic Tests Are Available?

A low white blood cell count (called leukopenia) is characteristic of this disease as is a low platelet count. But most cases of Colorado tick fever are definitively diagnosed by isolating the virus in the blood. A blood test can also be used to compare antibody levels during the acute and recovery phases of disease. A fourfold difference is considered proof of Colorado tick fever.

What Is the Treatment?

No antiviral therapy is available for Colorado tick fever, but your physician may prescribe medication to alleviate symptoms. Fortunately, a single episode of this disease is sufficient to confer immunity to future infection with the same viral strain.

TULAREMIA

Tularemia, also called rabbit fever or deerfly fever, is caused by the oval-shaped bacterium *Francisella tularensis.* The bacteria move into the bloodstream and travel throughout the body, eventually penetrating a community of immune cells that ordinarily protect the body against foreign substances. *Francisella tularensis* can survive for prolonged periods inside those cells.

Tularemia was first identified in 1911 by a researcher studying a plague-like illness in rodents. A year later, the causative agent was recovered from rodents in Tulare County, California, and was named *Bacterium tularensis.* The first human cases of the disease were reported 2 years later. In 1920, the bacterium was renamed *Francisella tularensis,* after a researcher named Francis who added to a growing storehouse of knowledge about the pathogen and the disease it causes. Ticks were shown to transmit *Francisella tularensis* in 1924. Today, approximately 150 cases of tularemia are voluntarily reported annually. The disease has occurred in every U.S. state, as well as in Canada, but the primary area of infection extends from the south-central parts of the United States to as far west as California.

How Is Tularemia Transmitted?

Numerous ticks transmit *Francisella tularensis*, including the black-legged, Western black-legged, lone star, Rocky Mountain wood, Pacific Coast, and American dog ticks.

Tick bites are responsible for about half the total cases of tularemia, mostly those that occur in the northern parts of the country in the spring and summer. In the southern United States, the disease is spread primarily by means other than ticks. Infection may also be transmitted in the bite of other blood-sucking arthropods, such as deerflies or horseflies, from touching infected animals or eating their meat—if it has not been adequately cooked—and by drinking contaminated water.

Muskrats and rabbits are the primary guardians of *Francisella tularensis* in the United States and Canada, but the source of tularemia in humans can also be found in more than one hundred other species of mammals, including cats, sheep, cattle, squirrels, deer, and coyote. Some birds, amphibians, and fish are also welcome hosts to the bacteria. Farmers, hunters, trappers, and military troops are most at risk for tularemia.

What Are the Symptoms?

Symptoms of infection generally begin within 3 to 5 days after exposure to *Francisella tularensis,* but some people notice symptoms within 1 day, and others do not become ill for 2 weeks.

The most characteristic sign of tularemia is painfully swollen lymph nodes, which can break through the skin, especially at the elbows and armpits, and develop into abscesses. Repeated spikes of severe fever occur in most patients, although some people only have prolonged low-grade fever. Chills and fatigue are also common. Other symptoms may include conjunctivitis or other eye problems, eye pain, a rash, and vomiting or nausea. The throat may hurt, and some people find it difficult to speak. Pneumonia or mild hepatitis have been reported, and the disease sometimes resembles typhoid fever. Serious complications, including respiratory distress, kidney failure, and heart or neurologic problems, can occur.

What Diagnostic Tests Are Available?

A blood test is used to compare antibody levels during the acute phase of disease with levels during recovery. A fourfold difference is considered proof of tularemia. Blood or tissue cultures are rarely taken because a special medium is required and because there is a risk of infecting laboratory personnel, who may inhale *Francisella tularensis* when it is released into the air.

What Is the Treatment?

Tularemia can be effectively treated with streptomycin, gentamicin, tetracycline, or chloramphenicol. A preventative vaccine is available for people whose jobs put them at high risk for tularemia. While it does not provide 100% protection, the vaccine will reduce disease severity.

RELAPSING FEVER

Tick-borne relapsing fevers occur throughout the world, except in a few areas of the Southwest Pacific, but in North America it is mainly found in the west and southwest. Reports of relapsing fever are sporadic, and the actual number of cases is not known. Infection is caused by at least three different close relatives of *Bb*. The first to be discovered was *Borrelia turicatae,* in 1933. Within a decade, researchers realized that different species of the *Ornithodoros* soft tick carried different species of the relapsing fever spirochetes.

Relapsing fever spirochetes are very sensitive to light and dehydration, are very slow to reproduce, and can survive freezing temperatures for months. They are maintained in nature in mammals including rodents, chipmunks, squirrels, rabbits, and mice. Lizards can also serve as tick hosts.

How Is Relapsing Fever Transmitted?

Ornithodoros (*O.*) ticks remain infected with relapsing fever spirochetes throughout their life cycles, and adult females can transmit infection to their eggs. Infection can be transmitted by the tick's saliva, so it is important to avoid touching the tick with bare hands. The infection can be transmitted within minutes after the start of the tick's feeding. Soft ticks typically inhabit remote, sheltered settings, including caves, mines, and wooden cabins. Infection sometimes occurs when a cabin is first opened for the season.

Borrelia hermsii is transmitted by the tick *O. hermsi. Borrelia turicatae* is transmitted by the tick *O. turicata.*

What Are the Symptoms?

Symptoms generally begin about 1 week after a tick bite, although there is a range of 4 to 18 days. Onset of disease is characterized by a sudden chill followed by a high fever and profuse sweating, giving rise to the term "shake, rattle, and roll chills." Typically, bouts of fever (as high as 105°F), last 2 to 9 days, followed by a week-long period when the bacteria are sequestered and there is no fever. This cycle generally occurs from three to ten times, usually for progressively shorter periods and with less severity.

Other early symptoms are headache and muscle and joint pain. Eventually, every organ of the body may be affected. About 25% of patients have a rash. Neurologic symptoms, including cranial nerve palsies, meningitis, and seizures, occur about 30% of the time. Inflammation of the eye, jaundice, and a nonproductive cough have all been reported. Patients may also experience difficulty breathing. Spleen, kidney, or heart problems can occur. Rarely, a patient will go into shock, and fatalities are occasionally reported.

What Diagnostic Tests Are Available?

These spirochetes float freely in the bloodstream when the fever associated with this disease is present. In 1996, a new test was developed at the National Institutes of Health's Rocky Mountain Laboratories that can detect all three relapsing fever organisms. The more traditional diagnostic approach has been to stain a blood sample taken during acute episodes of fever and look through a microscope for spirochetes. An antibody blood test is also used to confirm a relapsing fever diagnosis. A fourfold change in antibodies from the acute phase of disease to recovery is considered proof of relapsing fever.

Relapsing fever and Lyme disease tests can cross-react, so more than one test may be necessary to ensure an accurate diagnosis.

What Is the Treatment?

Tetracycline or doxycycline are usually prescribed to treat relapsing fever, although chloramphenicol is preferred for children and pregnant women. As in Lyme disease, a Jarisch-Herxheimer reaction, in which symptoms worsen as the medication begins to kill spirochetes, can occur within the first few hours of taking an antibiotic.

POWASSAN ENCEPHALITIS

Powassan encephalitis is caused by a tick-borne flavivirus, a type of virus that infects the brain. While it can be found across the United States, Canada, and Mexico, as well as in Russia, China, and southeast Asia, it is concentrated mostly in the eastern parts of North America. Only a few cases are reported each year.

How Is Powassan Encephalitis Transmitted?

The Rocky Mountain wood tick is the primary vector for the virus that spreads Powassan encephalitis. In 1996 researchers proved that the black-legged tick can transmit this disease. Several other species of the *Ixodes* tick family may also transmit the disease, but those that have been implicated rarely bite humans. Laboratory experiments have shown that infected goats shed the flavivirus in their milk, suggesting that there may also be a risk to anyone who drinks raw goat's milk.

Groundhogs, chipmunks, mice, foxes, and squirrels are able to maintain this infection in nature.

What Are the Symptoms?

The symptoms of Powassan encephalitis take several weeks to develop after a tick bite. The disease begins with flu-like symptoms, including fever, headache, fatigue, light sensitivity, pain behind the eyes, and muscle weakness. Eventually, a neurologic illness develops that is characterized by seizures, partial or complete paralysis, and olfactory hallucinations (you may smell something that is not there). Aphasia, which is a diminished or total inability to write, speak, or communicate, can occur, as can brain inflammation, dementia, and death.

What Diagnostic Tests Are Available?

A diagnosis of Powassan encephalitis can be confirmed with an antibody blood test, which must be administered by CDC officials. This test measures antibodies to the flavivirus during both the acute and recovery phases of disease. A fourfold increase in the recovery sample is considered proof that you had Powassan encephalitis.

What Is the Treatment?

No antiviral treatment is available for Powassan encephalitis, although you may be offered symptomatic relief, such as medicine to reduce fever.

TICK PARALYSIS

Tick paralysis is a potentially fatal reaction to a neurotoxin secreted in the saliva of a female tick at the late stages of her feeding. Tick paralysis has almost killed several children, including one in New York and one in Washington state in the past two years.

How Is Tick Paralysis Transmitted?

The adult female Rocky Mountain wood tick is the primary cause of tick paralysis, but the adult female American dog tick and possibly the black-legged and Western black-legged ticks can also transmit the neurotoxin to you as they feed. The lone star tick is also suspected, but this is controversial. On the West Coast, the Rocky Mountain wood tick is the culprit, while the American dog tick is usually implicated elsewhere in North America.

What Are the Symptoms?

Within 2 to 7 days after feeding begins, the tick's toxin begins to take effect. Paralysis begins in the lower body, especially the legs, and spreads within hours to the rest of the body, arms, and head. If the tick is not found and removed, this toxin can lead to respiratory failure and death.

What Is the Treatment?

If the tick is promptly removed, you will most likely recover within a few hours, but if nerve damage occurs, recovery may take longer. The bite of the Australian tick known as *Ixodes holocyclus* causes paralysis by a different mechanism and may be fatal even if the tick is removed.

8

HOW TO REMOVE A TICK

During 8 intensive years at the Lyme Disease Foundation, where I currently chair the board of directors, I have had some unusual telephone calls and heard a lot of bizarre stories. One of the stranger conversations I have had began with a caller who told me this story:

> I need some help. Yesterday my husband found a tick attached to his . . . well, let's just say his private parts. We don't have a family doctor, so I called the hospital, and a nurse suggested I coat the tick with petroleum jelly and see if it would slip off by itself.
>
> That's what we did, and then we kept a close watch on the tick throughout the day. We passed some of the time telling jokes. My son kept telling his dad that he seemed to have found a new blood brother. Someone else commented that he must have encountered a lawyer from the IRS. The jokes kept us chuckling, but the tick didn't show any signs of detaching.
>
> Several hours later, we called another hospital. This time, a doctor suggested we try burning the tick or pricking at it with a needle. I lit a match and tried to hold my hand as steady as possible, but my husband jerked away and got burned in the process. After we applied some ice to the sensitive area, I tried to dig the tick out with the needle, but all I managed to do was to cut into my poor husband's skin. Now, he won't let me near him. The tick has been attached for a day, my husband is in a foul mood, and I'm at the end of my rope. What should I do?

If it weren't for the obvious distress in the caller's voice, I admit that I would have taken this for a prank call. But I knew all too well how dangerous it could be to remove a tick improperly, especially in such a delicate area of the body, and I urged her to seek immediate medical help so that the tick could be properly removed with tweezers. I was concerned not only about the infections the tick might carry but also about the possibility that the creature would secrete a toxin that can be paralyzing.

Several months later, I got another call from the family. They had taken my advice and gone to a local walk-in clinic where the on-duty physician had removed the tick properly with tweezers. Unfortunately, the man was eventually diagnosed with Lyme disease. There is no doubt in my mind that delaying the tick removal, and then trying to pull it out improperly, significantly increased his chance of becoming ill.

KEY TO SAFETY: BE PROPER AND PROMPT

You can't fully appreciate the panic that occurs when you find a bloodsucking tick attached to your skin. I have seen very calm researchers reduced to sheer panic when they find a tick feeding on them. A bite never seems like a big deal—until it happens to you.

Assume, then, that finding an attached tick will cause you some anxiety, and if possible, ask someone else to remove it for you—your spouse, colleague, or a neighbor will do fine. However, once you understand the art of tick removal, it is not terribly difficult to do it yourself, should that be necessary. Be sure that your children know they should ask an adult to remove a tick from their bodies—they should never try to do it themselves or even ask another child to help. Remind your young ones that you expect to be called as soon as a tick is found.

Proper and prompt tick removal is the only way to reduce your chance of infection. That's probably the most important message I can deliver in this book. The longer an infected tick is attached, the greater the opportunity it has to feed; the longer it feeds, the greater the likelihood that a bite will lead to infection. Whether it is convenient or not, you should treat an attached tick as an immediate medical problem and seek treatment without delay. However, proper removal is even more important. At the Lyme Disease Foundation, we gave a great deal of consideration to whether to recommend *prompt* or *proper* tick removal as a first priority and voted overwhelmingly to advocate *proper* removal. If you don't know how to remove a tick properly, or you don't have the equipment needed to do it right, you may cause disease that would not otherwise have developed.

TICK REMOVAL EQUIPMENT

Removing a tick is not terribly difficult to do, although studies from tick experts in Australia have shown that the black-legged, Western black-legged, and lone star ticks are a bit trickier to remove than others because they insert their mouthparts more deeply into the skin. Here is what you need to remove an attached tick *properly* and *promptly.*

- **A Quiet Place:** It takes just a few seconds to remove a tick. It is best to go to a quiet, well-lit room, away from curious onlookers, so that both the person who has been bitten and the person who is removing the tick feel calm and unpressured.
- **Proper Tweezers:** Use only a blunt-tip, fine-point tweezers to remove an embedded tick. Avoid eyebrow tweezers, which can accidentally squeeze the tick, causing it to release its infectious juices into your wound. Tweezers with a sharp point are also a no-no because they may prick the tick, causing the release of infectious juices.

Finding the right tweezers is not easy. Unfortunately, there are a number of imaginative, but ineffective, products now being marketed to the public for the purpose of tick removal. Avoid empty promises and stick with the tweezer distributors I have listed in the appendix.

- **A Container:** A tick that has bitten you should be tested to determine whether it is carrying any infectious agents. To save a tick, you will need to have some sort of container on hand. A clean pill vial, a clear film container, or a Ziploc plastic bag works fine. Or you can place the tick on a piece of clear tape and fold it over, trapping the creature inside. That gives you a good view of the creature and may make it easier to identify.
- **Antiseptic:** Purchase a small box of individually wrapped packets soaked in an antiseptic, such as alcohol or betaine.
- **Adhesive Tape:** Buy a roll of white, waterproof adhesive tape about 1 inch wide at your local pharmacy so that you can label the tick container with information about where and when the bite occurred. Adhesive tape can also be used to remove a crawling tick if you find one.
- **Pen or Marker:** Waterproof ink allows you write on the container label without concern that it will run.

PROPER TICK REMOVAL

Bring your tweezers as close as possible to the point at which the tick is actually attached to your skin and grasp its mouthparts. (See Figures 8–1 and 8–2.) Then, pull the tick straight back, opposite from the direction in which the mouthparts are inserted, using steady, gentle pressure. Contrary to popular

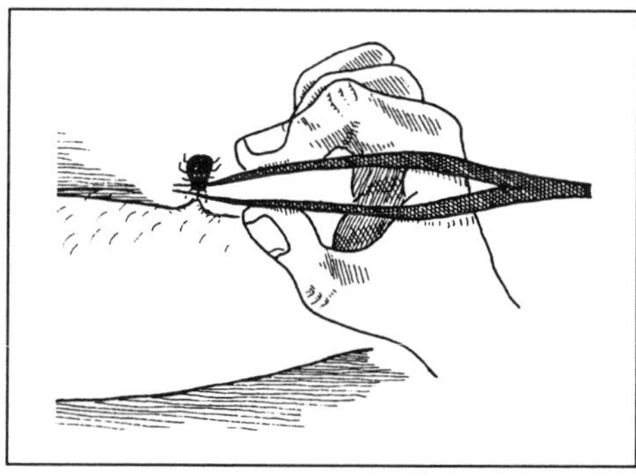

Figure 8–1
Proper tick removal. Reprinted by permission of Karen Vanderhoof-Forschner.

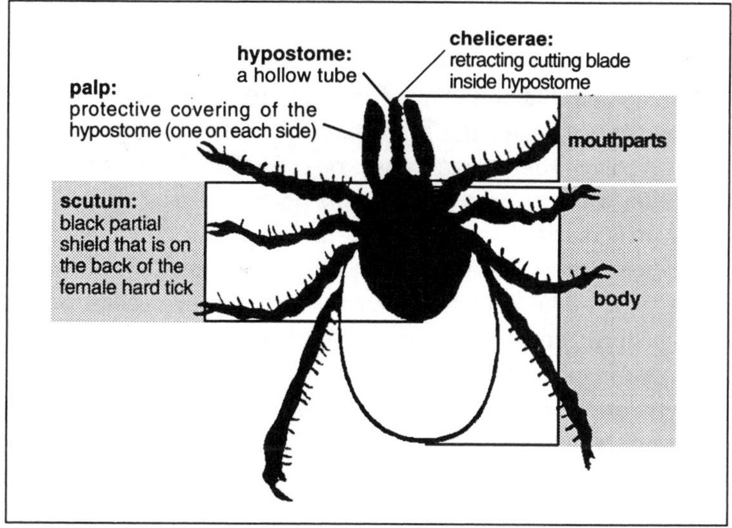

Figure 8–2. Adult female *Ixodes* tick. Reprinted by permission of Karen Vanderhoof-Forschner.

perceptions, ticks do not screw themselves into the skin and cannot, therefore, be "unscrewed."

Here are some important no-no's of tick removal:

- *Do not* twist the tick, which can cause the creature to break apart, leaving it partly lodged in your skin.
- *Do not* prick, crush, or burn the tick, which may cause it to salivate or regurgitate infected fluids.

- *Do not* try to smother the tick with products such as petroleum jelly, mineral oil, or nail polish. This is an ineffective technique because ticks can store enough oxygen to complete feeding.
- *Do not* wait for the tick to back out after feeding. Ticks do not behave that way.
- *Do not* use your fingers to remove the tick, which can lead to infection. It is almost always better to wait until you can secure the right tweezers, especially if it is only a matter of a few hours. If for some reason you must remove the tick with your fingers, use a tissue or leaf to avoid contact with potentially infected fluids.
- *Do not* crush the tick between your fingers after it has been removed. This alone could cause infection.

According to Glen Needham, writing in the June 1985 issue of *Pediatrics,* a tick may continue to salivate for a short time after it has been removed, allowing bacteria to enter the body through a break in the skin or if you subsequently touch your mouth, nose, or eyes. To avoid further risks, to either the person who has been bitten or the person helping to remove the tick, place the tick immediately in an appropriate container. Then, check your skin carefully to see if there is any evidence that the creature's mouthparts remain embedded, which could create a source of chronic irritation. If so, remove them with a sterilized needle, just as you would a splinter to eliminate any possibility of infection.

Both you and the person who removed the tick should then wash your hands thoroughly with a disinfectant soap. It is also important to disinfect the tweezers and clean the bite site with rubbing alcohol or betaine.

STORING OR DISPOSING OF TICKS

Open your tick container only when you are ready to place the tick inside. If you are sending it to a laboratory for analysis, you need to keep the tick alive. You can do so by adding two short blades of grass or a small piece of moistened tissue to the container vial, which provides the necessary moisture to sustain a tick for several days. If it takes you longer to find a testing site, place the tick vial in the refrigerator. Your state or local health department (generally listed in the government pages of the telephone book) or a nearby university can probably tell you where free tick identification or testing are available. The appendix of this book lists several other options. Be sure the laboratory you choose will test each tick for all of the pathogens known to be in the area where you were bitten. Tests generally cost between twenty-five and sixty dollars.

If you pull more than one tick from your body, you may find it difficult to use the same vial because ticks instinctively climb upward quite rapidly. As you try to add a new tick, the ones already inside may position themselves for an escape. If you don't have another container at the ready, try one of these strategies:

1. Rap the bottom of the vial several times against a hard surface to knock the tick down. Then, carefully open the top, drop the new tick inside, and close quickly.
2. Turn the vial upside down, which will cause the tick inside to climb upwards, to the bottom of the vial. Then, right the container, open, and quickly drop in the new tick. Again, close the container immediately.

If you decide to throw away the tick, wrap tape around the vial or place it in a Ziploc bag and toss it in the trash outside your home. Do not flush it down the toilet.

DOCUMENT AND WATCH

Write down key information about the occurrence of the tick bite on the container label. You should also write it in a permanent location in your home, perhaps on your family calendar, a datebook, or in immunization records. Or use the Personalized Medical Log I have provided in the appendix to help you accurately record your family's experiences with ticks.

Be sure to make note of the following information:

- Name of the person bitten
- Date of the bite
- Location of the bite on the body
- Estimated length of time the tick was attached
- The type of tick, if known
- How the tick was removed

You should contact your physician as soon as possible after a tick bite occurs. Many doctors treat patients with antibiotics immediately, while others prefer a watch-and-wait approach. Discuss the risks and benefits of both strategies.

If you and your doctor opt against immediate treatment, make a special effort to check yourself for rashes in front of a mirror every evening and keep notes of any physical complaints that occur in the 6 weeks following the bite. If you observe a rash, call your doctor's office for an immediate appointment. You should also mark the outside edge of the rash with a pen, keep the area dry, and watch to see whether it begins to enlarge. If so, take a photograph of it, because some rashes disappear quickly. If the rash is gone before you get to the doctor's office, bring the photograph with you. This photo should be kept as part of your permanent records.

MAKING YOUR OWN TICK KIT

I strongly recommend that you prepare a tick kit in advance and learn the art of proper tick removal before you actually need to do it. In fact, the best advice is to make several kits and keep them in convenient locations in the house or garage and in each of your automobiles. You might also want to make a kit for each member of your family to carry, especially if anyone is likely to stop for a walk or picnic in a local park or spend the weekend in the country.

Purchase a small zippered carrying pouch, about 4 by 6 inches, and insert these items:

- A blunt-tip, fine-point tweezers
- A 1-inch-wide roll of white adhesive tape
- Two empty, clear film containers
- Waterproof pen
- Tick repellent
- Index card to record relevant information

If you prefer to purchase a ready-made product, there is only one on the market that I can wholeheartedly recommend. It is called, appropriately enough, the Tick Kit, and it includes a proper tweezers, a vial to hold the tick, and an alcohol wipe. The Tick Kit is made of plastic, fits in your hand and even has a string attached so that you can conveniently wear it around your neck. The cost is just thirteen dollars.

Another popular product is a spring-loaded bow-shaped tweezer called the Tick Solution, which is now available as part of a kit from Scandinavian Natural Health & Beauty products, along with a case, a magnifying glass, and an alcohol wipe. Although several prominent physicians and entomologists swear by these tweezers, I've got some reservations about them and recommend that you file the tines to roughen them a bit before use.

For information on how to obtain these products, consult the appendix.

REMOVING UNATTACHED TICKS

If you notice a tick on your clothes, you can safely remove it with one of two methods that I've nicknamed to help you remember:

1. The Flick-It Method: Lightly flick the tick away with your finger or a stick.
2. The Stick-it Method: Make a tape-tick sandwich by carefully sticking a generous piece of tape over the tick. Lift the tape with the tick still attached and fold it over. Voilà, a tick sandwich. Be sure to discard it outside your home.

9

PERSONAL PROTECTION FOR YOU AND YOUR FAMILY

I can hear the music from the movie *Jaws* playing in my mind as I approach the potentially tick-infested woods behind my house: *da-da, da-da, da-da, da*. Then, the shrill violin sound in the shower scene of *Psycho* comes to mind. My own distaste for ticks has even led me to visualize the bugs taking out their forks and knives and getting ready for a tasty meal—with me as the appetizer!

My belief that prevention is the most important step in good health persuades me to incorporate the techniques of personal protection into my family's daily routine, and I urge you to do the same. Even with the best techniques of property management, which I describe in detail in chapter 10, it is almost impossible to keep your property entirely free of ticks and their hosts. The occasional deer still travels through my yard, and sometimes I see a mouse scurrying across my deck. Liberate your family by using appropriate protection strategies, and then go outside and enjoy yourself!

If your immune system is seriously compromised by any other disease, or if you are pregnant, avoid tick-infested areas altogether. Don't take chances, even if it means temporarily curtailing your activities—you may have to stop gardening and avoid all nature walks, for example. Tick-borne infections pose a serious health risk to a fetus, including the potential for miscarriage or stillbirth.

The rest of us don't have to go to such extremes, but we should recognize that we are taking a chance every time we venture into an area where ticks may linger awaiting a host. Since there will probably never be a magic bullet to protect you from all of the disease-causing agents spread by ticks, an ounce of prevention really *is* worth more than a pound of cure. What you do to prevent tick-borne infections can spell the difference between a healthy family and one plagued with chronic illness. Here are six important steps you need to take:

1. Educate your family about tick-bite prevention.
2. Dress properly.
3. Avoid contact with tick-infested vegetation.
4. Conduct tick checks any time you or a family member has been in a tick habitat.
5. Understand the risks and benefits of tick repellents and pesticides and make an educated decision about which ones to use.
6. Encourage local schools and businesses to take preventative action.

One final idea: Don't become totally consumed by the dangers of ticks. If you feel yourself becoming unduly fearful, ease your anxiety by listening to the relaxing sound of crickets and birds.

The six steps to prevention will now be discussed in detail.

EDUCATE YOUR FAMILY

Education may be the most important part of any prevention program designed to help you avoid tick bites. I can't emphasize this enough. Ignorance is not bliss. Tick-spread diseases can cause long-term chronic illness and permanent bodily harm. These illnesses can devastate a family's finances and destroy a marriage. Some of them can be life-threatening.

Fortunately, there are many tools available to educate your family and friends. The Lyme Disease Foundation produces and distributes a wide variety of videotapes and printed material for elementary school children, junior high students, and adults. I particularly recommend the *Dr. Ticked-Off* videotape for children. Many local or state libraries will allow you to borrow Lyme Disease Foundation videotapes or you can order copies from the foundation so that you can show them to family members at least once a year and to your houseguests as well. See the appendix for ordering information.

Once you are armed with the materials you need, sit down with your family and develop an understanding of what tick-spread diseases are and how to prevent them. Make the education a family event. You can begin teaching your children about ticks and how to avoid them by the age of two. Just remember that life can be scary to a little child, and watch your tone of voice and the words you use. Try not to sound unduly alarmed—we certainly don't want our children to become fearful of the outdoors.

While you should use age-appropriate language when you are talking about danger and illness, don't minimize your child's capacity to understand. I explain to my daughter that there are good bugs and bad bugs and that ticks are bad bugs that can make her sick. She also knows that her mother checks her for ticks every day to be sure they have stayed away. When you talk to your children, be reassuring, and make sure they understand that someone will take care of them if they ever get sick.

A two-year-old is also old enough to learn how to conduct a careful tick check, although you will want to double-check at least once a day until at least the age of six. Your children should know that you expect to be called right away if they find a tick attached to themselves.

DRESS PROPERLY

Proper dress is the number one rule that everyone in your family should learn to respect when heading into tick-infested areas. Here is what you need to know:

- Wear light-colored clothing, long sleeves, and long pants. Light-colored clothing allows you to see ticks more easily if they are crawling on your clothing. Long-sleeved shirts and long pants reduce the skin area exposed to ticks and keep ticks on the outside of your clothes.
- Tuck in your clothes properly. Tuck your shirt into your pants and your pants into your socks to prevent ticks from reaching your skin. By tucking your upper clothing underneath your lower clothing, ticks are forced to stay outside your clothes as they crawl upwards. If you are doing a lot of gardening or other outside work in tick-infested areas, use wide masking tape or duct tape to secure the place where your pants enter your socks and where your sleeves meet your wrists.
- Wear a hat. Keeping your head covered, especially when you are in areas with a lot of shrubbery, makes it harder for ticks to get lost in your hair or to reach your scalp. As a minimum level of protection, tie long hair back to prevent it from touching nearby brush.
- Wear tightly woven clothing. In a 1986 report of a study at the U.S. Department of Agriculture (USDA), research scientist Carl Schreck observed

that at all three stages of the life cycle, black-legged ticks had difficulty in clinging to relatively smooth, closely woven fabrics, such as cotton. Because the ticks were not firmly attached to the fabric, they often fell off clothing as a volunteer tester stood up.

- Wear shoes that cover your toes. Ticks can attach to your feet and ankles, so it is best to avoid open-toed shoes or sandals. At the very least, avoid bare feet.
- Never use animal tick or flea collars as human protection. This may sound a bit bizarre, but in the armed forces, where more military troops are felled by vector-borne diseases than by conflict itself, it is an issue of concern. Using these collars around your ankles is very dangerous and causes serious skin burns.
- Clean your clothes promptly. When you come inside from being in a tick-infested area, immediately put your clothes through the washer and dryer to kill any ticks that have decided to join you.

If You *Do Not* Dress Properly

In the heat of the summer, it may be impractical to wear long pants and long sleeves all day. And who wants to prevent the little ones from running around outside or playing in a sprinkler? Once the temperature soars much above 75°F, it is hard to resist the temptation to wear shorts and short-sleeved shirts, at least some of the time. If you do, be sure to redouble other prevention activities. Keep your shoes and socks on as often as possible and be rigorous about using repellents. You may want to make a particular effort to spray your socks with either tick-killing chemicals or repellent.

If you compromise on proper dress, *never compromise on your tick checks.* As you decrease your protection by reducing proper dress you must increase your tick checks to hourly and your naked tick-checks to several times a day.

Caution: Several years ago, a researcher went to a Connecticut nudist camp and surveyed the residents to see whether Lyme disease was a problem. On learning that almost no one had been infected, the excited physician wrote a letter that was published in the *New England Journal of Medicine* claiming that ticks don't bite fully exposed skin. The implication was that people would be safer if they wore a minimal amount of clothes, or none at all.

Unfortunately, experience has not borne this out. Most likely, the nudists simply had more opportunities to examine their own bodies and the bodies of their companions, making them more likely to spot and remove crawling ticks.

AVOID CONTACT WITH TICK-INFESTED VEGETATION

Ticks wait on brush—nymphs from about ground level to about 2 feet above ground, adults no higher than 3 feet above ground—until they can hitch a ride with a passing animal. Vegetation that may be infested with ticks includes the plantings around your home or apartment, the natural setting of local parks, the grass around shoreline beaches, and even the weedy overgrowth that encroaches on the walkways to stores and businesses.

To avoid unnecessary contact with this vegetation, walk on the center of paths, don't take shortcuts through the woods, and don't walk through brushy areas that lack clearly defined walkways. Also, avoid sitting directly on the ground. If you must sit on the ground, use a blanket that has been sprayed with tick-killing chemicals, as described later in this chapter.

INSPECT YOURSELF FOR TICKS

Frequent tick checks have become a way of life for my entire family, and I urge you to learn this habit. When you are in tick-infested areas, hourly checks are not too much. Otherwise, do a visual inspection of your clothing and exposed skin at least every 3 to 4 hours. Once you are back inside, take off all your clothes and examine your entire body in a private location. Use a mirror or ask a partner to help to make sure you don't miss any hard-to-see places. Remember, ticks are tiny—be on the alert for something that will look like a moving freckle.

You should also check your pets carefully, especially around the ears, eyes, or groin area. Pets have been known to bring both attached and unattached ticks inside the home. Attached ticks can cause disease in pets, while the pet may shake off an unattached tick, giving it an opportunity to choose some other member of your family for its next meal.

DECIDE WHICH TICK REPELLENTS AND PESTICIDES TO USE

Using Tick Repellents

A very important line of defense against ticks is to use tick repellents that have been approved by the federal Environmental Protection Agency (EPA). If the EPA is satisfied that a product repels ticks, the label will say so explicitly. Remember that no repellent is 100% effective against all ticks, so never eliminate your careful tick checks.

Even so, there are many challenges to choosing the right product. Different species of ticks are repelled at different rates, and a chemical that repels one kind of tick very effectively might work less well against another. Unfortunately, the black-legged tick that carries *Bb* is not as easy to repel as other ticks, according to Carl Schreck, the USDA scientist who is fondly known as the "father of repellents."

Certain products are only effective for a very limited periods of time. There may also be questions about the effectiveness of a repellent at different stages of the tick's life—what works against an adult may be less effective against the nymph or larva of the same species. The same product may also have different effects when applied to clothing, rather than to the skin, yet many scientific studies fail to analyze both factors. Weaker repellents may be safer or more pleasant to use, but they also may not be as effective. Some repellents are quicker to evaporate, more readily absorbed into the skin, or more likely to be washed away with rain or sweat, all of which could have an impact on effectiveness. Further complicating the picture is that ticks may be less interested in biting once they have been removed from their natural geographic location and shipped to a laboratory for the testing of repellents.

If a product merely says it is effective against insects, you cannot assume you are safe. One day, I sat on a park bench next to a woman who was applying a popular children's repellent to her little boy. She saw the Lyme Disease Foundation logo on my shirt and leaned over and said, "See, I'm protecting my son from ticks." I asked her to check the repellent label to see if it mentioned ticks. She was startled to discover there was not a word about them. Her child was completely unprotected against tick bites. The lesson: Read the label carefully.

There are questions and controversies about most products available, and making a decision is not easy. Although I've got my own preferences, as described below, I've also tried to give you the information you need to make an informed decision about other repellent products.

Avon's Skin-So-Soft Mosquito, Flea and Deer Tick Repellent

If I were to recommend a single repellent that provides protection against Lyme ticks, for everyday use I would advise you to use Avon's Skin-So-Soft Mosquito, Flea and Deer Tick Repellent. It is especially good for children and pregnant women because it is an all-natural blend that has no toxicity. I do not receive any compensation for endorsing this product, but I find it very effective. I did have the opportunity to review the scientific studies relating to this product, with the help of an entomologist advisor to the Lyme Disease Foundation, and we were both impressed with the care and attention that had been given to the scientific details.

While citronella-based products don't generally work well against ticks, Avon's product uses oil of citronella as its active ingredient. The product is a combination of sunscreen, moisturizer, and EPA-approved repellent. The product also has several "secret" ingredients that apparently help boost its ability to repel ticks and mosquitos. Two scents are available—the Skin-So-Soft authentic fragrance and the Herbal Fresh scent. You have a choice of SPF 15, a partial sun block, or SPF 30, a full sun block. You should reapply the product every 80 minutes if you go swimming.

Read the product label carefully before purchasing it. The Mosquito, Flea and Deer Tick Repellent is a completely different formulation from Avon's other Skin-So-Soft products and bears the same name for marketing purposes only. This product is safe enough for me to allow my three-year-old daughter to apply it by herself, although I watch to make sure she doesn't put the cream into her mouth. It can also safely be used by pregnant women. Some people are also using it on horses, but Avon makes no claims about safety for animals.

Deet-Based Products

Deet (N,N-diethyl-m-toluamide) is the main active ingredient in most popular tick repellents. Deet was first synthesized in the 1950s by the USDA and has been commercially available since 1956. Studies have proven that it effectively repels a wide variety of arthropods, including ticks, fleas, chiggers, and mosquitoes, and the chemical is currently used by tens, perhaps hundreds, of millions of people every year. Products containing deet go by such familiar names as *Off!*, Cutter, Muskol, and Ben's. They are widely available and marketed in many different forms, from aerosol sprays to towelettes, and contain deet at varying concentrations.

A longer lasting product has been developed by the 3-M Corporation in response to a Department of Defense request. The odorless repellent contains 33% deet in a long-lasting, controlled-release formulation and is now used by U.S. troops around the world. The 3-M product, called Ultrathon, uses a slowly evaporating polymer that reduces both skin absorption and deet evaporation and has proven valuable against a number of tick species. While 3-M makes this product only for the armed forces, the company has licensed it to Amway, which markets it for direct distribution to the public as "Hour Guard." The product repels arthropods for about 12 hours, although its effectiveness gradually tapers off.

Does Deet Work? Deet's effectiveness against ticks has been the source of some controversy. Research findings are mixed and highlight the challenge of measuring effectiveness. We know that the product certainly has great value against some ticks. In a 1983 study that appeared in the *Journal of Economic Entomology,* researchers showed that products with 20% deet were found to be 85% effective against the lone star tick and 94% effective against the American dog tick. The black-legged and Western black-legged ticks, which carry the Lyme disease–causing *Bb,* were not included in this study. Nor did the study show how long the repellent was effective or whether it worked equally well at each of the tick's life stages.

In another study, this one authored by Carl Schreck and published in the *Journal of Medical Entomology* in 1986, researchers showed that when 20% deet was sprayed on clothing it repelled 86% of the black-legged and Western black-legged ticks, and that 30% deet repelled 92% of the ticks. But four years after that publication, repellent companies voluntarily removed a controversial substance named R-11, a chemical that had been linked to cancer, from their repellents. Some researchers speculated that R-11 had some action to repel ticks, so its removal raises questions about the product's effectiveness.

Another Schreck study, published in the *Journal of the American Mosquito Control Association* in 1995, helped answer some of the questions about the value of deet in low concentrations against *Bb*-carrying ticks. Researchers tested the ability of deet and 28 other chemicals to repel the black-legged and lone star ticks and found that the chemicals were dramatically more effective against the lone star tick. In fact, a compound with 25% deet properly applied to the skin provided complete protection against the nymphal black-legged tick *for only 10 minutes*—by the time you put the repellent on, grab your cap, and walk outside, it has lost much of its effectiveness! This study suggests that if you are going to use a deet-based product—and some people may want to do so because it is also effective against mosquitos and other insects—you should use the higher concentrations to increase your protection from *Bb*-carrying ticks. Certainly there doesn't seem to be much sense in using the standard children's formulation, which has only a 10% deet concentration.

Is Deet Safe? Unfortunately, the use of deet products raises troubling questions about safety. There have been a number of reports of adverse reactions to the chemical, which are not dose related. Although millions of people have used deet with no problems at all, documented reactions include skin rashes, fever, blistering, severe headaches, hepatitis, behavioral changes, mood disturbances such as crying and irritability, nausea, insomnia, poor coordination, and muscle spasms. There have also been incidents in which the user became confused, spoke in slurred language, suffered a seizure, or fell into a coma. Tragically, deet has also been associated with the death of several children.

These reactions can be traced to the fact that deet is a neurotoxin that is absorbed through the skin, circulates throughout the body, and can be recovered from the urine. In standard deet-based products, studies show that as much as 56% of the chemical penetrates the skin and 17% is absorbed into the circulatory system, according to Theodore F. Tsai in *Arboviral Infections: General Considerations for Prevention, Diagnosis, and Treatment in Travellers.* Half of the absorbed dose is then excreted in the urine over the following 5 days.

Some of the problems with deet clearly result from improper use—if you eat this chemical or lather it on your skin excessively, without washing between applications, the chances of getting sick obviously increase. Even with proper usage, however, some people, especially children, are sensitive to the chemical. Some medical groups recommend limiting the deet concentration of products used on children to a maximum of 10%. However, this does not solve the problem because the deet-related side effects are not concentration related. Worse yet, 10% deet probably leaves children *unprotected* against tick bites. Some experts have suggested that the alcohol (ethanol) content of deet-based products may increase absorption into the body.

Because of deet's potentially serious side effects, some health departments have warned the public not to use products containing more than 30%, to use them sparingly, and to wash when returning inside. In 1995, the New York State Legislature banned the sale of all repellents containing deet concentrations exceeding 30%. However, state officials were not of one mind. The state health commissioner, Barbara De Buono, objected, saying that higher concentrations are more effective and should be allowed in light of the public health threat of Lyme disease. The state Department of Environmental Conservation agreed and issued an emergency regulation allowing the use of higher concentration products, and that's where things stand today.

While research continues, I believe it is wise to listen to the health experts who privately suggest that deet may be especially risky to pregnant women, children under four, and people in poor health. My recommendation is that pregnant women avoid tick-infested areas altogether, eliminating both the need to use a repellent and any lingering risk of infection. I also advise against deet-based products in children under four.

Using Deet Properly

Most people use deet without experiencing any side effects, but it is wise to apply the product sparingly to exposed skin and vital to follow package directions. If you want to use a deet product, be sure to respect these cautions:

- Do not apply deet-based products near your lips or eyes.
- Apply repellent to your hand, then use your hands to spread it on your face.
- Never apply deet repellents to young children's hands, as they may then touch their eyes and mouths.
- Apply deet to exposed skin only. Applying the repellent to skin covered by clothing can result in chafing and lead to more serious skin irritation.
- Never apply deet over rashes or on sunburned, wounded, or otherwise broken skin, because injured skin absorbs more of the chemical.
- Avoid excessive or prolonged use. Do not use every day for the whole summer.
- Wash the repellent off when you return inside.
- Keep repellent containers out of the reach of children. Accidental ingestion could kill your child.

You may also want to spray repellent on your clothes, but you should do it cautiously, as some fabric can be damaged. Deet is safe on cotton, wool, and nylon, but Spandex, rayon, and acetate don't stand up very well to deet, and it dissolves plastic and vinyl. A 1993 *Consumer Reports* study of repellents also indicated that when deet is applied to polyester acrylic cloth, the fabric becomes highly flammable.

Use Pesticides on Your Clothes and Belongings, as Needed

Certain chemical pesticides can be applied to your clothing, before you put it on, and to other articles that you take into tick-infested areas but should never be applied directly to exposed skin. One of the most effective pesticides is permethrin, marketed as Duranon Tick Repellant (formerly Permanone), which is a synthetic imitation of natural substances found in chrysanthemums and other flowers, called pyrethins. Duranon has .5% permethrin as its active ingredient and is sold in most lawn and garden shops. Troops in the armed services apply this to their uniforms and find that it lasts through several launderings.

Permethrin kills ticks as they walk across treated fabric because it is neurotoxic, causing disorientation and death to almost any arthropod within minutes of coming into contact with it. The chemical binds to fabric and will not stain, has a pleasant odor, resists degradation by light or heat, and is nearly insoluble in water. A commercial permethrin product known as Perma-kill is also available for dogs and is effective against ticks, fleas, and lice.

To use permethrin safely, hold the can about 6 inches from your clothing and spray the fabric until it looks moist. Spray both sides of the clothing and let the fabric dry for at least 2 hours before you wear it. One 6-ounce can will spray two complete sets of clothes. A good idea is to treat one special rugged outfit and then plan to wear that whenever you work outside.

You may also want to spray your picnic blanket, and if you are camping, spray the walls, ceiling, and floor of the tent in which you will be sleeping. Don't forget to plan ahead.

Other important tips for the safe use of permethrin:

- Do not apply to the clothes as you wear them.
- To avoid breathing the vapors, apply the spray outside.
- Do not touch the clothing until the material is completely dry.
- Reapply permethrin after the fifth washing.

Remember, too, that permethrin has no value as a repellent, which means that a tick can bite your hand next to your sleeve even if your clothes have been thoroughly treated. Because the pesticide is poorly absorbed and rapidly inactivated, it has a very low level of toxicity to mammals. Occasionally, people report local irritations. Although the chemical is registered as a possible carcinogen, there is no reason to believe it is toxic in its dry form.

START A SCHOOL OR BUSINESS PREVENTION PROGRAM

What the Schools Can Do

When it comes to getting the word out about tick-borne diseases and how to protect yourself from them, there is work to be done not only in the home but also at work, in the schools, in your church, synagogue, or mosque, and in civic associations. Anyplace people gather is the right place to spread the word about the dangers of ticks and what can be done to prevent illness.

School-age children and teenagers unfortunately have many chances to be bitten by ticks during recess, gym, and after-school sports activities, so schools need to develop a program for managing ticks and preventing bites. If you are a parent or teacher, think about approaching the PTA or talk to school health officials about the problem. Find out what is being done, and if necessary, organize your peers to press for action. In my hometown of Tolland, Connecticut, a fabulous safety program is in place, even though we have not yet had a major tick problem. Our school officials believe it is better to be safe than sorry, which is the absolutely best approach.

Here are three of the preventive steps your school should be taking:

Step 1: Education

- Creative teaching techniques allow tick education to be incorporated into the curriculum of almost any subject. Science and health classes at every grade level are obvious places to begin, but many other approaches, such as using a Spanish-language videotape about Lyme disease as a tool to teach Spanish, also work well.
- Handouts and other written information about tick-borne diseases should be sent home with the students. The Lyme Disease Foundation can provide many of these materials. You may also want to suggest that articles about tick prevention be published in the school newspaper or PTA newsletter.

- Parents should be asked to give informed consent when a school trip takes students into potentially tick-infested areas.

Step 2: Tick Checks

- When students come in from recess, they should routinely line up in a circle to conduct quick visual tick checks on each other. If a tick is found, the student should go to the school nurse to have it removed properly, and the parents should be immediately informed. A proper tick-removal kit should be available in the school, and personnel should be trained to use it. If there is no school nurse on duty, the parents should be called and urged to take the child to the doctor immediately. When possible, school health officials should place the tick in a vial and send it home with the student so the parents can seek appropriate medical follow-up.

Step 3: School Property Management

- Techniques of property management should be applied to reduce the tick population. In particular, brush should be cut back around playing fields, picnic areas, and anyplace else the children gather. See chapter 10 for more details.

What Business Can Do

At the workplace there is also a responsibility to educate and protect employees at all levels, as well as any customers that come on the premises. Some businesspeople have been afraid to talk openly about tick-borne diseases, apparently believing it could lower property values, increase lawsuits, or scare away customers. My response: *"Wake-up!"* Just the opposite is probably true—most likely, responsible education and heightened awareness will decrease liability, increase health, and encourage the responsible use of the outdoors by newly empowered people.

Some workers come into direct contact with tick-infested areas, including police officers, people who install or maintain utility and telephone lines, anyone who handles dogs or other animals, construction workers, postal delivery employees, park rangers, grounds maintenance workers, and railroad workers. Even the people who collect the garbage may be exposed to ticks. But if you think you are safe because you have a desk job, think again. Corporate golf outings, company picnics, fund-raising events, tennis challenges, and company retreats all can place you outside and in the company of ticks.

Employees are not the only ones in danger. Customers and clients may also come in contact with pathogen-transmitting ticks at commercial sites. You could be bitten at a restaurant that provides picnic tables, at a riding stable, or at a swim club. Even corporate parks that open their gates to the public for concerts or other special events may inadvertently expose people to infected ticks.

Whether you own a business or work in one, take some responsibility for helping to put a creative and well-organized tick-bite prevention and education program in place. Everyone stands to gain. Employees and customers are less likely to face the consequences of illness, while employers will benefit from reduced health care and disability costs, less absenteeism, and lower rates of employee turnover associated with long-term illness. The risks of liability may also be lessened if a jury can be persuaded that a business took appropriate action to lessen risks.

Step 1: Education

Anyone who comes onto a business property—whether employees, family members, or customers—should be educated about ticks. Education is simple and very inexpensive, especially compared to the tens of thousands of dollars it can cost for the health care of a single employee affected by Lyme disease. There are many creative ways to incorporate education into communications vehicles that already exist. Here are five:

1. Include information about Lyme disease and other tick-borne disorders as pay-envelope stuffers or in the company newsletter.
2. Invite guest speakers to make presentations to workers during safety meetings, luncheon gatherings, or departmental meetings.
3. Host a community presentation at your workplace—it is a great way to build community relationships as well as to provide information.
4. Establish a health care lending library with a wide variety of information about tick-spread disease, including pamphlets, newsletters, books, and videotapes. Make this book the first addition to the new library, and contact the Lyme Disease Foundation for other materials.
5. Post signs and distribute flyers to warn employees and visitors that ticks might be in the area and to remind people to check themselves thoroughly.

Step 2: A Shared Responsibility

Once everyone understands the risks of tick-borne diseases and how to avoid them, it becomes a lot easier to make safe practices part of the daily routine. All the techniques of personal protection described in this chapter are just as important in the workplace as they are at home. Employees who spend a lot of time outdoors should be provided with repellents and possibly permethrin-impregnated overalls if their work will take them into potentially tick-infested wooded or brushy areas.

A full-length mirror should be provided in the employee bathrooms or changing rooms to encourage full-body tick checks. This is also an ideal place to post information about conducting a proper inspection and how to obtain assistance with tick removal. The work site should have a full tick-removal kit, or at least a pair of tweezers, and everyone should know where to go on site to have the tick removed properly.

Businesses should also establish tick-free zones where people are most likely to gather, such as picnic areas, outdoor recreation facilities, shortcuts around the building, and walking paths. Overhanging branches should be trimmed, any brush adjacent to walkways should be cut away, and the lawn should be mowed regularly. Employers may also want to consider the risks and benefits of using tick-killing chemicals on the land. For more details, refer to the next chapter, which explains proper property management.

10

MANAGING YOUR PROPERTY

More ticks are to be found in woodlands and forests, including the woods in your own backyard, than in any other type of terrain, according to the Connecticut Agricultural Experiment Station. The second most popular gathering place? Your own lawn. The lawn's edge is especially popular with mice and other small animals, which are the preferred hosts of nymphal ticks, particularly if the property borders woods or is marked by a stone wall.

Fortunately, there are many steps you can take to reduce the chances of contracting a tick-borne disease anywhere in your yard. The main goal of property modification is to create tick-free zones in some of the most well-traveled areas of your property and to reduce the number of ticks everywhere. The costs are modest, but a commitment of time is required. In the long run, I assure you it is an investment worth making. You will reduce your own chances of getting disease, protect family members and friends, and even reduce the risk of liability if a visitor contracts a tick-spread infection on your property.

Unfortunately, ticks have no consistently effective natural predators at any point in their life cycle. Scientists were initially hopeful that either chalcid wasps, tiny and short-lived insects that lay their eggs in nymphal ticks, or guinea fowl, which eat and peck at almost everything, might help control the tick population, but studies show they don't make a significant difference. Fire ants may be more effective, but they are very selective—according to the Texas Department of Health, they won't eat the brown dog tick but they can clear an area of all its lone star ticks.

Despite the limitations, there is a great deal you can do to make your land less tick-friendly and more people-friendly, so get busy. Most changes cost little and are relatively simple.

COLLECTING TICKS AT HOME AND NEARBY

Two tick-gathering methods will help you determine whether there are infected ticks on your property, and a third is designed to give you more general information about the presence of ticks in your community. Before you embark on this project, however, please take this warning to heart: Collecting ticks can be a dangerous business. Even properly dressed entomololgists can't always avoid disease. In fact, one estimate from Westchester Medical Center in Valhalla, New York, is that as many as one-third of Lyme disease researchers who go out into the field to collect ticks eventually contact tick-spread disorders.

You need to assemble several items for tick collection:

- Fine-point, blunt-tipped tweezers
- A container, such as a clean pill jar or clear film case, with a fresh blade of grass inside
- White adhesive tape to use as a label
- Waterproof ballpoint pen
- A flag or dry ice

Remember to dress carefully—review chapter 9 on techniques for personal protection. In particular, I emphasize the need to use wide masking tape or duct tape where your pants enter your socks and where your sleeves meet your wrist. Be especially vigilant about checking your nude body for ticks when you are finished.

The Flagging Method

The best time of year to flag (see Figure 10–1) for black-legged nymphal ticks is in April and May; September is best for adult ticks. However, there are regional variations, so call your local or state health department, a nearby college's science or entomology department, or the local agriculture station to ask about the best time for flagging.

Make a flag, using a piece of white corduroy or flannel fabric measuring about 2 by 3 feet. Staple the shorter side of the cloth to a broomstick or the handle of a shovel so that it looks like a flag on a pole. Alternatively, you can make a "drag" where you staple the cloth to the center of the pole and attach a long rope to both ends of the pole. The small nap in the material causes the

Figure 10–1. Flagging. Reprinted by permission of Karen Vanderhoof-Forschner.

ticks to attach to the fabric. Do not use terry cloth because it tends to catch on the shrubbery and allows ticks to become embedded in the knobs.

In the early or mid-morning, slowly drag your flag on the ground over a 10-foot-square area, poking it several times into any brushy undergrowth between the lawn and the woods or adjacent property, which is generally the best place to find ticks. After running the flag over the designated area, slowly turn it over and examine the side that had been touching the ground. As you inspect the cloth for ticks, be careful not to let your body or your clothing come in contact with the cloth.

If you find a tick, use tweezers to pull it off the fabric and place it in your container.

The Dry Ice Method

Using dry ice is not as effective as flagging for black-legged or Western black-legged ticks but works very well for lone star ticks. As dry ice evaporates, it emits carbon dioxide, which is a lure to ticks. Purchase a block of dry ice from your local liquor store, ice cream parlor, or supermarket, taking care to handle it appropriately as it can burn your skin.

In the early evening, place a 4-foot square of white corduroy or flannel fabric in the area that you want to test for ticks. Then place the dry ice on top of this cloth. Inspect the cloth for ticks at least three times during the day, in the morning, late afternoon, and early evening.

Again, use tweezers to pull any ticks from the fabric and place them in your container.

Checking the Kill

Many states have weighing or inspection stations to which hunters bring animals they have killed during the hunting season. Some tick collectors go to these weigh stations and collect attached ticks from the bodies of the tick's natural hosts. Call your state game licensing commission for locations and dates. While this can't actually tell you whether infected ticks have invaded your property, it is the best method for proving that they inhabit the local area.

VEGETATION MANAGEMENT

Many of the strategies for discouraging ticks in your yard have two general goals:

1. To reduce the humidity on the ground.
2. To move the vegetation that allows ticks to reach their hosts farther from your home.

Ticks are highly vulnerable to dehydration. A study by the Connecticut Agricultural Experiment Station showed that lawns that get a lot of sunlight, which lowers the humidity, have almost no ticks, even if there are abundant ticks in the adjacent woods. On the other hand, plants like pachysandra and ivy are ideal ground cover for ticks, because they hold high concentrations of humidity, protect ticks from the drying effect of sun exposure, and provide shelter for mice and other hosts.

Here are some techniques that will help:

- Prune or eliminate trees and overhanging branches and clear away brush so that more sun and air reach the soil.
- Move the forest, and the ticks it may contain, by extending the areas of open lawn around your house.
- Remove leaves, branches, and other vegetative litter.
- Mow the grass regularly to reduce ground-level humidity. By one estimate, mowing alone can reduce the tick population by 70%.
- If there are fields as part of your property, keep them trimmed to about 3 inches high.
- Let the grass dry thoroughly between waterings. Better yet, don't do any watering at all.
- Widen trails and walkways so that they are at least 6 feet wide and make extra efforts to keep them clear of brush.
- Move your outdoors clothesline to an open area of your lawn so that draped sheets and blankets do not touch the brush.
- Move children's play areas out of the woods.

HOST MANAGEMENT

Another part of property management involves modifying your land to make it less attractive to the animals that serve as tick hosts. On their own, ticks are sedentary bugs that travel less than 10 feet throughout their life. When they attach to mice, they travel an additional 20 feet. However, when they attach to deer and other larger hosts, they can travel several miles more. Birds are the most efficient transportation form of all, especially during migration seasons, when they can carry ticks across continents. With all of the hitchhiking possibilities, it is no wonder that infected ticks keep appearing in new places.

Reduce your risks by discouraging their hosts. These techniques work well:

- Eliminate bird-feeders and bird baths. Let wildlife feed themselves. It may sound cruel, but birds have a plentiful supply of natural food and can easily find places to wash themselves.
- Eliminate deer salt licks.
- Erect 10-foot-high fencing around your property, which discourages

deer and may reduce the tick larva population by as much as 97%. You'll also eliminate about half the population of nymphal ticks and nearly three-quarters of the adults. It is best to electrify your fence with a small charge and then use bait, such as peanut butter, to lure the deer to the fence. The mild electric shock the deer will experience makes a lasting impression, without harming the animals, and they will avoid the fence in the future.

- Use repellents. Check with your local lawn and garden center to find out what commercial products are most effective against the deer population. Usually, products made with eggs or blood products are best, but unfortunately these sometimes attract other mammals, such as raccoons.
- Remove wood piles and stone walls, which provide homes for mice, chipmunks, and other small animals that may serve as tick hosts.
- Remove garbage, move garbage cans farther away from your house, and make sure the lids fit securely.
- Modify your landscape design. Decrease the variety of soft plants that deer like to eat, including roses, rhododendrons, and forsythia, and increase the plants that deer dislike, especially those with thorns, a pungent taste, or woody stalks. Examples are the Colorado spruce, holly bushes, boxwoods, and marigolds. Talk to the experts at your local nursery for additional suggestions.
- Acquire a dog, preferably one with a loud bark. This may seem odd, but dogs will chase away, or scare off, many animals and birds. Trust me—this suggestion works!
- Never move livestock or pets to a new area until you are certain the animals are tick-free. Otherwise, you may introduce ticks to an area where there have not been any.
- Never take in stray or wild animals. The danger of rabies is well-established, but the chance that you will befriend a tick-infested host is sometimes forgotten.

I can't recommend the strategy of trying to kill tick hosts, such as mice and deer, since it seems to make little difference. In one study, a deer herd was reduced by 70% without having a significant impact on the tick population—ticks simply latch on to other hosts.

KILLING TICKS

In the 1940s, it was recognized that chemicals worked very effectively to kill a wide variety of arthropods, including mosquitoes and ticks. Within a decade, efforts to reduce the human and animal diseases they spread had led to the widespread use of synthetic insecticides. Unfortunately, it soon became apparent that these chemicals had a much wider impact than had been recognized—along with their target populations, they also killed harmless bugs, fish, and birds, causing significant harm to interlaced ecosystems.

Over the next 30 years, chemicals were modified to be safer. Today's tick-killing chemicals, known as acaricides, biodegrade in a matter of days or weeks. That's a big step forward, but they remain toxic to birds and fish, and sometimes to other animals as well. Concern also lingers about their adverse effects on humans as the chemicals make their way into the water supply and the food cycle. In *Chemical Exposure and Diseases,* author Joan Sherman argues that chlorpyrifos, also known by the brand name Dursban, which is highly effective against ticks, overloads the human nervous system and causes dizziness, nausea, and muscle pain. Chlorpyrifos is also being investigated for its links to an unusual pattern of birth defects involving the eyes, ears, palate, teeth, heart, feet, nipples, genitalia, and brain. Similar birth defects have been reported from the use of other pesticides as well.

Nonetheless, acaricides are legal, and they can significantly reduce all stages of ticks. It is up to you to weigh the risks and benefits to your family and decide whether to use them. Along with chlorpyrifos, cyfluthrin (brand name, Tempo) and carbaryl (brand name, Sevin) are quite effective. Researchers disagree about whether the liquid or granular form of these pesticides is most effective. Some say liquid application with a high-pressure spray kills ticks more swiftly, while others are convinced that granules penetrate dense brush more effectively without leaving residue on the vegetation. The liquid form is less costly but slightly more complicated to use. The more expensive granules can be easily applied with a fertilizer spreader or a hand cyclone spreader. Regardless of the product you choose, be sure to follow package directions carefully and consult with the experts at your local lawn and garden shop.

Insecticides should be applied initially in late May or early June in the northern United States and late February to early March in the south to kill nymphs and larvae. A single application in the late spring can reduce the nymph population by 100% within 3 days. The larvae population is also greatly reduced—by anywhere from 50% to 90%, according to one study. A second application between September and early October kills the remaining larvae and most or all of the emerging adults. By killing the ticks that would otherwise lay eggs, you significantly increase your protection for the following year.

Another approach takes advantage of the fact that mice are scavengers that line their nests with any goodies they can find. One inventive individual put cotton balls treated with the pesticide permethrin in cardboard tubes, such as the ones you find inside a roll of paper towels. He then distributed the tubes around the edges of the property where tick-infected mice would find them and steal the cotton balls for their nests. The goal, of course, was to have the permethrin kill any tick on any of the mice sharing the nest. If mice are the primary host for all the ticks on your property, this method works fairly well, although you may still have to contend with visits from neighboring mice. More significantly, most ticks feed on more than one type of animal, so you aren't likely to eliminate your entire problem.

Research into other pesticides continues. In one Texas study, deer are being fed corn specially treated with the pesticide Ivermectin (a product also given to dogs to prevent heartworm), the theory being that as the pesticide circulates in the deer's blood, it will be ingested by feeding ticks and will kill them. So far, this appears to be very effective in controlling tick-spread diseases, especially cattle tick fever, which has plagued the area since the 1950s. However, the deer themselves cannot be safely eaten for 2 months after ingesting the dose of pesticide, suggesting possible limitations for a more widespread use of this technique. More work in this complicated area is needed.

Don't try burning the vegetation around your house as a means of killing ticks. Although the tick population may be temporarily reduced, burning actually eliminates tall overgrowth and allows the excessive proliferation of new low-height shrubbery . . . producing ideal conditions for mice and tick habitats. This results in a larger mouse population the following year, giving ticks access to even more hosts. Farmers have known about this phenomenon for years. Burning also creates pollution and carries a fire risk that can threaten your home.

11

THE PROMISE OF VACCINES

One of the greatest medical challenges facing researchers today is to prevent infectious diseases altogether. While treatments for many diseases have advanced considerably, they come at enormous cost, both in terms of medical expenses and human illness. Combine this with the emergence of bacteria that have become resistant to standard antibiotic treatment and the importance of avoiding infection altogether becomes apparent.

Vaccines are one of the best prevention methods available. By exposing your body to inactive bacteria, or a piece of it, a vaccine prompts your body to generate an antibody response. Then, if you should encounter a live version of the same pathogen, you will have the antibodies in your system ready to incapacitate or kill it.

A rumor has been circulating that a vaccine for Lyme disease exists and has been in widespread use in Europe for many years. Sadly, this is not true. The only tick-borne disorder in humans for which an effective vaccine does exist is encephalitis, a viral infection of the central nervous system that is found in Central Europe, Scandinavia, Russia, the Far East, and the countries that formerly made up the USSR. Even this vaccine is not 100% effective, and it has some side effects. Several Lyme disease vaccines have been developed for dogs and these are described in more detail in chapter 12.

But the rush is on in the pharmaceutical industry to be the first with a human vaccine against Lyme disease. The potential sales could reach $200 to $300 million in the first few years after approval from the Food and Drug Administration (FDA), according to industry analysts. Any subsequent improvements on the vaccine will also bring significant revenue to the manufacturer. I certainly don't begrudge the vaccine makers a profitable return on their investment—for those of us who have endured the burdens of this disease, it would be an extraordinary breakthrough.

THE CHALLENGES

Unfortunately, the search for an effective vaccine will likely be slow, tedious, and filled with hurdles. The current challenges are to find a piece of the Lyme bacterium that is common to all strains and that is exposed to your immune system as soon as it enters your system. An effective vaccine must also evoke an antibody response that will kill all of the bacteria without confusing your body to the extent that it attacks its own cells or mounts some other dangerous reaction.

A further obstacle to vaccine development is that almost all pharmaceutical research and testing is enormously time-consuming and costly. According to a 1996 report by the congressional Office of Technology Assessment, it costs $500 million and takes 12 years for a company to move one new medicine from the laboratory to the pharmacy shelves. The FDA oversees the vaccine testing process, which has three major steps:

1. **Pre-clinical Testing:** For a period of 3 to 4 years, a manufacturer will conduct laboratory and animal studies to demonstrate safety, biological activity, and success rate. Once this is completed, the company applies to the FDA for an Investigational New Drug Application that allows human clinical trials to begin.
2. **Clinical Trials:** Over a period of about 6 years, the company will study a vaccine in three phases, enrolling increasingly larger numbers of people. During Phase 1, a small number of healthy human volunteers test the vaccine so that appropriate dosages and safety issues can be studied. During Phase 2, several hundred human volunteers who may be at risk for the disease test the vaccine for effectiveness and side effects. In Phase 3, as many as several thousand human volunteers at risk for the disease must complete these trials to confirm a vaccine's effectiveness and to monitor for long-term adverse reactions.
3. **Evaluation:** Once all of the human trials are complete, a company evaluates its data and submits them to the FDA. On average, another 18 months elapse after the company requests approval until the product becomes commercially available.

THE TRIALS UNDER WAY

Thousands of healthy volunteers are part of vaccine trials being sponsored by at least four different manufacturers. My hat is off to these courageous pioneers whose efforts are crucial to moving science forward. Even if the current trials don't produce a viable vaccine, they will undoubtedly contribute to scientific knowledge and pave the way for other advances.

Young people under the age of 17 and pregnant or nursing mothers are excluded from current studies as are those who have Lyme disease or other serious medical conditions. In most of the trials, half the study population is given an active vaccine while the other half receives a placebo, which is a pharmacologically inactive product used as a tool of comparison. Researchers are watching carefully to see if they can detect significant differences between the two groups.

All of the current human vaccine trials are devoted to the membrane that covers the Lyme bacterium, which contains the outer surface protein A (OspA). OspA initially seemed promising because it is present in many of the known Lyme bacteria strains in the United States and it evokes an antibody response that can immobilize the bacteria. Originally, it was believed to be the structure the immune system first encountered as the bacteria entered the body.

Unfortunately, the National Institutes of Health's own scientists have shown that the Lyme disease bacterium is one step ahead of the vaccine developers. Once a tick infected with *Bb* attaches itself to a warm-blooded mammal and begins to feed, it apparently stops producing OspA and begins to make OspC instead. OspC thus becomes the first foreign substance to provoke an immune response, which helps explain why the body apparently fails to make antibodies to OspA, or to still another protein, OspB, early in the infection process.

Other limitations to the concept of a vaccine targeted at the OspA is that it is not widely conserved around the world, meaning that it is not found in a significant portion of the Lyme bacteria. While OspA antibody inhibits the growth of about 82% of the strains in the United States, it is of use only in about 14% of non-U.S. strains.

Clearly, more work will need to be done. Nonetheless, the current trials are an important beginning. These are the studies under way:

Pasteur-Merieux Connaught Laboratories

In 1991, Pasteur-Merieux Connaught became the first manufacturer to launch human Lyme disease vaccine trials. Its product was developed by Alan Barbour, M.D., who is a scientific advisor to the Lyme Disease Foundation, and his colleagues at the University of Texas. Dr. Barbour patented the OspA vaccine to one *Bb* strain in 1988 and subsequently sold the rights to Pasteur-Merieux Connaught Laboratories. In early trials, the vaccinated groups were able to immobilize the Lyme bacterium without any apparent side effects. Now, larger populations are being followed, and the results are promising.

SmithKline Beecham

This vaccine, developed by researchers at Yale University School of Medicine, involves the OspA of another *Bb* strain. In preclinical studies, researchers were startled to observe that when an infected tick fed on a vaccinated animal, the animal's blood entered the tick's body and killed the infection before it ever reached the animal. An important study question is whether the same thing occurs in humans. By 1995, eleven thousand volunteers were enrolled in 32 centers across the country. The results of this testing are also promising.

Pfizer Central Research

Pfizer Central Research is developing an OspA vaccine that can be delivered either by injection or orally. The antibodies are delivered using a killed strain of the *Salmonella* bacterium; sounds odd, but this is an established delivery system. In trials with mice, the oral vaccine was effective 80% of the time.

MedImmune

MedImmune has sponsored an OspA vaccine trial since 1994, but it has also been moving its research in a new direction. In 1993, researchers at Texas A&M University discovered a *Bb* protein that binds with a protein called decorin, which is found in nearly all human skin and joint tissue. The possibility of incorporating the decorin-binding protein (DBP) into a vaccine during the early stage of infection sparked optimism within the research community. Preliminary animal experiments prove that DBP inhibits the growth of all strains in the United States and about half of those beyond our borders, showing that the protein is widely conserved. And most importantly, the DBP protein is exposed to your immune system when bacteria enter the body.

Subsequent experiments on mice demonstrated that a DBP vaccine effectively prevents disease if it is given within 4 days after exposure to the *Bb*, suggesting the vaccine may be useful in treatment, as well as in prevention. A vaccine will not be effective in human beings, however, until researchers are able to purify the protein fully and determine if the immune system response it prompts is sufficient to kill all of the disease-causing bacteria.

THE FUTURE

Great advances have been made, but science still has a long way to go. Until we have a more definitive test to determine who has active Lyme disease infection and who does not, it will be very challenging to measure how well a vaccine works. We also need to resolve a long-standing debate about people who have continuing symptoms of illness. Some scientists believe that these people are persistently infected with *Bb;* others think the problem is the failure of the immune system to shut down properly after eradicating the infection. The answer has important implications for vaccine development because a vaccine could worsen the medical consequences of an overactive immune system. Early reports from ongoing clinical trials, however, suggest that persisting symptoms come from persisting infection. More knowledge about the process by which a tick feeds on, and infects, its hosts will also enhance the process of developing a fully protective vaccine.

In the next year or two, new Lyme disease vaccine trials will certainly be launched. The possibility of an OspC vaccine is being explored, and researchers are also interested in developing a "cocktail" vaccine that would incorporate several components of the *Bb.* Future tick vaccines are likely to deliver their active ingredients via nasal spray or an oral medication, rather than an injection, because animal studies suggest that this may be the best way to produce a potent, long-lasting, and systemic antibody response.

Researchers are also interested in developing a vaccine that will cause a tick to rapidly detach or die before it can transmit disease. And I expect to see more attention given to an all-purpose tick vaccine, rather than to one targeted exclusively at Lyme disease. Given the number of tick-spread diseases that threaten us, government-funded research in this area would be highly appropriate.

A controversial question is who should be eligible for a vaccine once it is developed. Some gatekeepers to the public health have suggested that the vaccine only be given to people who are at high risk, such as people who work in parks or live in areas of very high incidence. My position is different. I believe that fairness dictates that everyone should have the right to decide whether to be vaccinated against Lyme disease. I hope it is not long before a viable vaccine forces us to debate that issue.

12

THE RISKS TO YOUR ANIMALS

As the struggle to develop a human vaccine continues, and efforts to understand, diagnose, and treat Lyme disease and other tick-borne disorders move ahead, one population suffering from these infections has been overlooked for too long. Your beloved household pets, as well as any livestock you may raise, are also at risk.

In my own household, we had to confront the problem of Lyme disease in our two dogs, Ruddy and Dancer, and our three cats, Mozart, Daphne, and Puffy. When the dogs began having trouble walking and seemed feverish, I quickly made a veterinarian appointment. After checking them over, the vet said it was nothing more than a cold, but back home, they couldn't shake their symptoms. The symptoms became so severe that we had to take them to the emergency room at the veterinary hospital. Soon, they were limping, and Ruddy became unable to follow the commands he had learned at obedience school. To make a yearlong story short, we went back to the vet, who began to suspect Lyme disease and sent us home with antibiotics. A Lyme disease test confirmed the diagnosis, but Ruddy and Dancer never recovered.

At the same time, my cats were showing signs of illness. I had noticed them trying to pull ticks off their bodies, and soon afterwards, they were unable to walk and whimpered in pain whenever I tried to pick them up. Eventually, I mentioned the possibility of Lyme disease to my veterinarian, but he initially dismissed it, saying it did not affect cats. A year later, I insisted on a Lyme disease test, and sure enough, the results were positive. The vet prescribed antibiotics, but one by one, our beloved cats had to be put down due to serious Lyme-related problems. Two of them had developed serious eye problems, and the autopsy I requested revealed *Bb* bacteria in their bad eyes.

There are many reasons to be concerned about Lyme disease, babesiosis, and ehrlichiosis in pets and livestock. If infected ticks are in the environment, they may easily infect more than one animal. Also, animals are sentinels for human disease. If your pet has been exposed to infected ticks, or diagnosed with Lyme disease, every other member of your family may be at risk and should be medically evaluated. Even if no one else becomes sick, disease affects the entire household—medicines must be administered, time may be lost from work to secure veterinary care, and medical bills can be startlingly high.

SOURCES OF ANIMAL INFECTION

Cats, dogs, horses, and cows do not develop immunity to most tick-borne diseases after an initial infection, so they may become infected repeatedly. Domestic animals can be exposed to infected ticks almost anywhere. While an animal that hunts or is free to roam in the fields may be at special risk, just walking around the yard in some parts of the country is enough to expose your pet to pathogens for Lyme disease and other illnesses.

Animals can also carry infected ticks from one region to another. I used to watch horse shows that took place on a vacant overgrown field behind my Connecticut house and wondered if any of the owners knew that half the ticks on that land were infected with *Bb*. At the end of the show, some of them were undoubtedly attached and feeding and were likely to be transported to new areas at the next horse show, sometimes thousands of miles away.

Animals may also become infected with Lyme disease when their noses, mouths, eyes, or open wounds in the skin come into contact with pathogen-laden urine, although *Bb* does not survive once the urine has dried. Infection is most likely to be a problem in situations in which animals are confined or are in very close contact. Most female domestic animals are also able to pass infection on to their fetuses, and cats can theoretically develop Lyme disease after eating infected mice.

Most animals are not susceptible to all the tick-borne disorders that can affect humans, but it has already been well established that *Bb* does cause illness in domestic animals. Investigations into a newly discovered spirochete, temporarily named *Borrelia lonestari* sp. nov. and apparently transmitted by the lone star tick, may also threaten your animals. In addition, at least five species of *Ehrlichia* are known to cause disease in animals:

1. *Ehrlichia canis:* The often-fatal infection caused by this parasite was first

discovered in military working dogs in Vietnam in the mid-1960s. It is called canine monocytic ehrlichiosis, or tropical canine pancytopenia, and is often associated with severe bleeding.

2. *Ehrlichia ewingii:* This parasite causes canine granulocytic ehrlichiosis, which primarily infects the granulocyte cells of dogs.
3. *Ehrlichia phagocytophila:* This parasite infects cows and sheep with tick-borne fever.
4. *Ehrlichia equii:* The parasite causes an infection named equine ehrlichiosis, which targets the monocytic cells of horses and dogs and was recognized in 1996 as the cause of human granulocytic ehrlichiosis.
5. *Ehrlichia risticii:* This parasite causes Potomic horse fever.

DIAGNOSIS

A responsible veterinarian will conduct a thorough physical examination of your pet to see if its signs are consistent with Lyme disease. In addition, a blood sample will probably be taken in order to run a *Bb* antibody test. Because Lyme disease mimics other disorders in animals, just as it does in people, the vet may order a number of additional laboratory tests to rule in or out other conditions.

According to a Cornell University study, false positives, which suggest your animal has Lyme disease when it does not, are rare. False negatives, which fail to find any Lyme disease antibodies in your pet's blood, are more common. They may occur either because the animal has not yet mounted an immune system response to the infection or because a persistent infection has compromised its immune system, preventing it from responding appropriately.

Researchers disagree about the significance of positive Lyme tests if your animal has no noticeable signs of infection. Many veterinarians believe that a positive test indicates active infection. Steven Barthold, a Yale University researcher, injected animals with dead Lyme disease bacteria and found that it had no association with illness, lending support to the argument that persisting illness is due to persisting infection. Max Appel, a doctor of veterinary medicine (D.V.M.) at Cornell University School of Veterinary Medicine, supports this view, observing that an animal produces high levels of antibodies when an infection persists and that the immune response diminishes once the pathogen is killed.

Sandy Bushmich, Ph.D., D.V.M., a professor at the School of Pathobiology at the University of Connecticut, suggests that a dog that tests positive for the

Lyme disease bacteria but shows no signs of disease may be infected by a less virulent *Bb* strain. Still others believe that a few animals can be exposed to *Bb* and develop antibodies to it without ever being infected. Julie Rawlings, M.P.H., an epizootiologist at the Texas Department of Health, believes an animal may produce antibodies after eating a smaller infected animal, such as a mouse. It is theoretically possible that the larger animal does not become infected but is exposed to pieces of *Bb* in the mouse and develops measurable antibodies.

TREATMENT

The antibiotics of choice to treat infected animals are usually members of the tetracycline or penicillin families. Most animals improve dramatically after about 2 to 4 weeks. The Jarisch-Herxheimer reaction, in which the illness appears to worsen as treatment begins, may occur in animals, just as it does in people.

Several aspects of animal treatment are controversial. Questions about active infection make it difficult to know just how long treatment should last, and there is no consensus among veterinarians. A 1996 study by Max Appel found that the majority of dogs treated with the standard four weeks of antibiotics continued to be infected despite their antibody tests turning from positive to negative. There is also a debate about the use of intravenous (IV) antibiotics. Some researchers believe IV medications, because of their cost, should be given to larger animals (horses, cows) or those with commercial value (e.g., show dogs).

Another area of uncertainty is whether a pet that tests positive for *Bb* antibodies but has no signs of illness should receive treatment. My answer is yes, although others disagree. Some veterinarians worry that treating every animal that tests positive will involve vast numbers of animals. For example, as many as 89% of dogs have *Bb* antibodies in areas where Lyme disease is prevalent. I think the risks of overtreatment are worth taking because tissue and joint damage occur slowly, and you might not notice them until your pet is really suffering. Also, you have the greatest chance of reversing the damage of Lyme disease and restoring your pet to good health with early treatment. Still another reason to treat asymptomatic infection is to reduce the spread of disease. A feeding tick can carry bacteria from one infected animal to many others—by killing the bacteria in your pet, you can help break that cycle.

Once your pet begins a course of antibiotics, you should continue treatment faithfully for the time recommended by your veterinarian. Many owners are tempted to stop treatment when the animal appears to be getting better, but that allows any bacteria that linger to continue reproducing and possibly to mutate to a more resistant form. As an extra measure of precaution, you should keep infected pets apart from uninfected pets, at least for the first 2 weeks of antibiotic treatment. This type of quiet time might also be appreciated by the sick pet.

DISEASE PROFILE IN SPECIFIC ANIMALS

As with people, the signs and symptoms of Lyme disease in animals can range from very mild to severe. The range of medical problems also differs somewhat from one species to another.

Cats

Cats that are infected with *Bb* may have fever, difficulty in walking, and a loss of appetite. Fatigue is also common, and eye infection can occur (Figure 12–1). Sometimes, the cat's breathing is difficult or peculiar, and heart prob-

Figure 12–1. Cat with *Borrelia burgdorferi* in the eye. Reprinted by permission of Karen Vanderhoof-Forschner.

lems may be reported. On the other hand, signs of disease are often quite mild, and many owners may not even realize that they have an infected pet. Research about Lyme disease in cats is very limited, partly because researchers didn't realize until recently that cats could contract it.

Most cats respond well to 4 weeks of amoxicillin, which is generally added to food in liquid form.

Dogs

Only about 5% to 10% of dogs infected with *Bb* actually show signs of illness. Those that do may be fatigued and have shifting lameness and difficulty walking because of stiff, painful, or swollen joints (Figure 12–2). Fever is sometimes present, the lymph nodes may be swollen, and skin rashes may occur. Many dogs also eat poorly, lose weight, and become hypersensitive to touch. Signs of neurologic involvement include mood swings, depression, head tilt, aggressive behavior, seizures, confusion, palsies, progressive paralysis, and more. Eye problems occur causing conjunctivitis, retinal detachments, and inflammation of all areas of the eye. Heart damage can occur resulting in irregular heartbeats, a slow heartbeat, and an enlarged heart. Liver

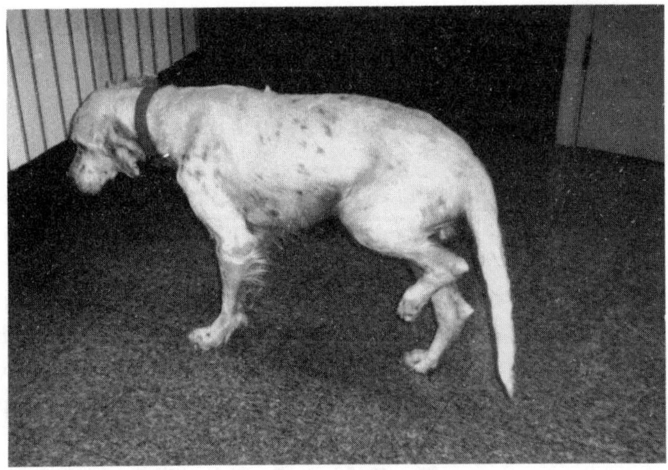

Figure 12–2. Dog hurting with joint paint. Reprinted by permission of E. Masters, M.D.

problems sometimes occur. Where kidney damage occurs, which may be apparent if there is blood or excess protein in the urine, the dog unfortunately has a poor prognosis. The Lyme disease bacterium has been cultured from a dog's muscles, joints, skin, spinal fluid, eyes, kidneys, lungs, heart, and the lining of the brain, which explains why multiple body systems may be affected by disease.

The precise clinical picture depends on the dog's age and on the life stage and the species of the tick that transmitted infection. For example, puppies are more likely to become lame than are adult dogs. And dogs bitten by adult ticks are more likely to develop serious problems and to test positive for Lyme disease antibodies than dogs bitten by nymphal ticks, possibly because adult ticks may be able to transmit more bacteria.

Lyme disease antibodies cannot be found in laboratory tests for 4 to 6 weeks after a tick bite, and false negatives are fairly common. Additionally, signs of infection often do not become apparent for 2 to 5 months, making early treatment almost impossible. When treatment is prescribed, oral amoxicillin or oral doxycycline are likely to be used for at least 2 to 4 weeks, although recent findings suggest that the length of treatment should perhaps be longer. If signs of illness return after treatment, additional antibiotics are usually recommended.

Horses

A recent study in New England found that only 10% of the horses that tested positive for Lyme disease antibodies showed obvious signs of disease. While some owners may be able to detect subtle changes in a horse's behavior, a great deal of infection is probably being overlooked. This is particularly troubling, because as many as one-quarter of the horses in New England may have *Bb* antibodies.

An infected horse generally has no or low-grade fever. Your horse may find it difficult to walk and may suffer from stiffness or swelling in the joints, general weakness, laminitis, and abnormal muscle movements. Because horses, especially those that carry riders or are used for jumping, are subject to injuries, other causes for these signs need to be ruled out. A more distinctive sign of Lyme disease is golf ball–size cellulitis, a tissue inflammation that sometimes occurs at the site of the tick bite. Loss of appetite, profuse sweating, and eye disorders, including moon blindness and loss of vision, may also occur. If there is kidney damage, it is usually noticeable as blood in the urine or fluid retention. Heart damage, leading to an irregular heartbeat, and liver damage have also been reported.

Sometimes neurological infection suggests Lyme disease. Numerous signs include a tilting of the head, difficulty swallowing, depression, aimless wandering, glazed eyes, paralysis of the tail, and difficulty following commands. It is not uncommon for horses to be hypersensitive to touch, which you may notice if the animal resists your efforts to groom it or place a saddle on its back. Spontaneous abortion and birth defects in the colts of infected mares have been documented.

Horse owners may be advised to give their animals daily injections of intramuscular penicillin or intravenous oxytetracycline. Oral tetracycline can also be used, but it sometimes causes intestinal disturbances, such as colitis or diarrhea. Response to antibiotics may be the best indicator that a horse has Lyme disease.

Cattle

Infected cattle may not be recognized because they often show no noticeable signs of illness or only mild behavioral changes. Among the signs that may be apparent are the erythema migrans rash, which may appear on the udder; swollen lymph glands, fever, difficulty walking, joint swelling, and stiffness, especially in the hocks. The legs or hoof area may be inflamed or stiff, and weight loss and eye problems, including moon blindness and loss of vision, sometimes occur. An infected cow may also produce less milk and may experience reproduction problems, including infertility, spontaneous abortion, and stillbirths.

In cattle that have been diagnosed with Lyme disease, the *Bb* spirochetes have been found in colostrum, milk, blood, the synovial fluid that surrounds the joints, the lungs, the liver, and spontaneously aborted fetal tissue. The special risk of Lyme disease to cattle is that an entire herd may become infected. Cattle showing obvious signs of infection are more likely than asymptomatic cattle to shed bacteria in their urine.

Fortunately, cattle treated with antibiotics, including penicillin, oxytetracycline, and ceftofur usually recover rapidly, although some animals may relapse and require retreatment. Although pasteurization kills the *Bb* organisms in infected milk, the milk of a cow known to have Lyme disease is not used until several weeks after signs of infection are gone. Meat is safe if it is cooked thoroughly at a temperature of at least 160°F.

PREVENTING TICK BITES

Some basic precautions should be taken to reduce the possibility that your pets or other livestock will be bitten by ticks. Even if your animals are housebound, preventative measures are in order. Although the black-legged and Western black-legged ticks that carry *Bb* don't live indoors, other pathogen-carrying ticks do.

Dog Vaccines

If you have a dog, you may want to consider one of the three vaccines against Lyme disease that are now on the market. Talk to your veterinarian about the risks and benefits of dog vaccines but remember that you still need to check your animals for ticks and to use tick repellents or pesticides. Vaccines are not 100% protective, and your animal faces risks from other serious tick-borne diseases. Precautions are also needed to prevent your dog from bringing live ticks onto your property and into your house.

The first two vaccines to be developed used all of the pieces of a killed Lyme bacterium. In 1989, Russ Johnson, Ph.D., of the University of Minnesota, who is a former member of the Lyme Disease Foundation and still a loyal friend, patented the first whole-cell dog vaccine. The patent was subsequently sold to Fort Dodge Laboratories, which made some improvements and now markets it as Bacterin. The vaccine has proved fairly safe and confers some protection against a number of *Bb* strains. A second vaccine, called Galaxy Lyme, is the first use of a vaccine cocktail made from two different killed strains of the Lyme bacterium. Since it entered the marketplace in 1994, it has been fairly successful in preventing disease.

Unfortunately, whole-cell killed vaccines have a remote chance of actually infecting your dog and also carry some risk of inducing an adverse reaction. To improve vaccine options, Pasteur-Merieux Connaught introduced a recombinant dog vaccine, which uses a piece of the *Bb* bacterium. A similar product for use in humans is currently in clinical trials.

Your dog should be tested for *Bb* before being vaccinated. A vaccine does not eradicate existing infection, so you may first have to treat the animal with antibiotics.

Other Prevention Strategies

Here are some other techniques to protect your dogs or livestock:

- Learn about tick-killing chemicals that are safe for animal use. Ticks that attach to dogs, cats, and horses can be killed with sprays containing permethrin or pyrethrins. However, Lloyd Miller, a doctor of veterinary medicine (D.V.M.) in Troy, New York, warns pet owners to consult a veterinarian before using tick-killing chemicals because some of them can be very dangerous. For example, certain insecticides can be fatal to cats.
- If you decide to use a tick-killing chemical, follow the manufacturer's instructions carefully and watch for any signs that your animal is reacting to the chemicals. Don't use a product designed for one type of animal on another species, and never apply more than one type of repellent to a single animal—mixing different chemicals can make your animal very sick or even be lethal. To protect yourself, always wear rubber gloves when using chemicals.
- Treat your dog's kennel and pet bedding with permethrin to eliminate ticks. Carpets can also be treated with a formulation of pyrethrin or allethrin. You may not be able to see the larval and nymphal ticks that have settled into cracks and crevices, but they may be there, near your pets and preparing to feed.
- Use the property management techniques described in chapter 10 to reduce your pet's exposure to ticks.
- Take special precautions when your pet enters tick habitats. Use repellents and tick collars and try not to let your pet roam free in the woods or fields.
- Examine your animals closely for embedded ticks, especially after they have been outside for any length of time at all. Start at the animal's head and move your hand down its body, reaching through the fur to feel for lumps or bits of dried blood that may signal a tick. Pay special attention to the ears, belly, and groin area, and don't forget to look all the way down the legs.
- Remove ticks properly and promptly to avoid further risks to your pet or yourself (for details on tick removal, see chapter 8). Once you have removed the tick, place it in a small container and label it with the date, your name, address, and telephone number, and the type of pet that was bitten. Contact your veterinarian to find out how the tick can be identified and tested for infection.

- Seek veterinary attention if your animal shows any signs of disease. Remember, the sooner an infection is diagnosed, the easier it is to treat.
- Have all of your pets evaluated if one animal has Lyme disease. While testing all animals in tick-infected areas is probably impractical, I think it is reasonable to test everyone in a home that has experienced Lyme disease.
- If you raise farm animals, make sure they are free of ticks and diseases when shifting one animal or herd from one area to another. Otherwise, you may bring infection to areas that are currently disease-free.
- Ask friends who are planning a visit to leave their pets at home to lessen the risk of introducing new infections.

Appendix I

TIMELINE: THE HISTORY OF LYME DISEASE

I have included references for written sources of information. To increase readability, I have left blank those items that involved personal communication or where I was at the event.

1883
German physician Alfred Buchwald describes a degenerative skin condition lasting 16 years in one patient, which he names *diffuse idiopathic skin atrophy*. This is the first record of what we now know as *acrodermatitis chronica atrophicans* (ACA), which is a late Lyme disease (LD) skin condition that occurs most frequently in Europe.

Source: Buchwald, A. (1883) "Ein Fall von diffuser idiopathischer Haut-Atrophie." *Arch Dermatol Syph* 10:553–556. Referenced in Weber/Burgdorfer (1993), chapter by Weber/Pfister.

1884
Ixodes ticks attached to a cat are collected and preserved in Hungary. More than a century later, they are found to have been infected with *Borrelia burgdorferi (Bb)* and become the oldest known record of infection.

Source: Matuschka et al. (1995)

1886
ACA is described as a "crumpled cigarette paper skin," a descriptive term that is still used today.

Source: Pospelow, A. (1886) "Cas d'une atrophie idiopathique de la peau." *Annales de Dermatol,* 7:505–510. Referenced in Weber/Burgdorfer (1993), chapter by Weber/Pfister.

1888
Ixodes ticks attached to a fox are collected and preserved in Austria. More than a century later, they are found to have been infected with *Bb*.

Source: Matuschka et al. (1995)

1894
A researcher from a Massachusetts museum collects and preserves white-footed mice. One hundred years later, *Bb* is found in their pelts, and this is recognized as the oldest known record of infection in the United States.

Source: Marshall et al. (1994)

1895
The early inflammatory phase of what is now called ACA is described and called *erythromelie*.

Source: Pick, P. J. (1895) "Über eine neue Krankheit erythromelie." *Verh Ges Dtsch Naturf 66,* Verlag Wien, 1894, II, 366. Leipzig 1895. Referenced in Scrimenti (1995).

1895
The first U.S. cases of what is now called ACA are described by an American dermatologist. The cases are seen among immigrant populations and resemble those described by Pick.

Source: Bronson, E.B. (1895) "A case of symmetric cutaneous atrophy of the extremities." *J Cutan Dis,* 13:1–10. Referenced in Scrimenti (1995).

1895
The multiple phases of what is now called ACA are described by an American dermatologist, who noted the initial presentation as a purplish discoloration of the skin, followed later by skin atrophy.

Source: Elliot, G.T. (1895) "A case of idiopathic atrophy of the skin." *J Cutan Dis,* 13:152. Referenced in Scrimenti (1995).

1901
Three stages of what is now called ACA—an early inflammatory stage, a late inflammatory stage with atrophy, and a final atrophic stage without inflammation—are described by Krzysztalowicz.

Source: Finger, E. and M. Oppenheim. (1910) "Die Hautatrophien." *Deuticke, Wien.* Referenced in Weber/Burgdorfer (1993), chapter by Weber/Pfister.

1902
Physicians Karl Herxheimer and Kuno Hartmann introduce the name *acrodermatitis chronica atrophicans,* which is later also referred to as *ACA Herxheimer*. They describe an early inflammatory stage and a later atrophy stage to the disease.

Source: Herxheimer, K. and K. Hartman. (1902) "Über acrodermatitis chronica atrophicans." *Arch Dermatol Syph,* 61:57–76, 255–300. Referenced in Weber/Burgdorfer (1993), chapter by Weber/Pfister.

1905
Herxheimer expands the description of ACA to include fibrous nodules.

Source: Herxheimer, K. and W. Schmidt. (1910) "Über strangförmige Neubildungen bei acrodermatitis chronica atrophicans." *Arch Dermatol Syph,* 105:145–168. Referenced in Weber/Burgdorfer (1993), chapter by Weber/Pfister.

1909
At a meeting of the Swedish Society of Dermatology, Swedish physician Arvid Afzelius describes a ringlike lesion of approximately 1 inch that expands outwards and is clear in the center. The rash, which is believed to come from the bite of an *Ixodes* tick, is today known as erythema migrans (EM). It has also been referred to as erythema chronicum migrans (ECM), erythema chronicum migrans Afzelius, erythema chronicum Lipschütz, and more.

Source: Afzelius, A. (1910) "Verhandlungen der Dermatologischen Gesellschaft zu Stockholm." *Arch Dermatol Syph,* 101:404. Referenced in Weber/Burgdorfer (1993), chapter by Weber/Pfister.

1910
Wilhelm Balban, an Austrian physician, describes a bluish-red circular rash that follows an insect bite and lasted 10 days. Today, we recognize this as another example of an EM rash.

Source: Balban, W. (1910) "Erythema annulare enstanden durch Insektenstiche." *Arch Dermatol Syph,* 105:423–430. Referenced in Weber/Burgdorfer (1993), chapter by Weber/Pfister.

1910
A report is published on ACA that includes patient characteristics, scleroderma-like changes, and skin atrophy.

Source: Finger, E. and M. Oppenheim. (1910) "Die Hautatrophien." *Deuticke, Wien.* Referenced in Weber/Burgdorfer (1993), chapter by Weber/Pfister.

1911
Swiss researcher Jean Louis Burckhardt identifies the first lymphocytoma to be preceded by a skin rash referred to as *erythematous plaque* and probably what we now call the erythema migrans rash. The lymphocytoma is attributed to chronic inflammation.

Source: Burckhardt, J. L. (1911) "Zur Frage der Follikel und Keimzentrenbildung in der Haut." *Frankf Z Pathol,* 6:352–359. Referenced in Weber/Burgdorfer (1993), chapter by Weber/Pfister.

1913
B. Lipschütz, an Austrian dermatologist, describes a long-lasting rash that he names *erythema chronica migrans*. The name is later shortened to erythema migrans. A controversy ensues as to whether the rash is caused by a pathogen or a toxin.

Source: Lipschütz, B. (1913) "Über eine seltene Erythemform (erythema cronicum migrans)." *Arch Dermatol Syph,* 118:349–356. Referenced in Weber/Burgdorfer (1993), chapter by Weber/Pfister.

1915
For the first time cases of lymphocytoma-like lesions and surrounding EM rashes are described in European children.

Source: Kerl, W. (1915) "Erythema chronicum migrans." *Arch Dermatol Syph,* 119:301. Referenced in Weber/Burgdorfer (1993), chapter by Weber/Pfister.

1920
For the first time, lymphocytomas are associated with tick bites, a link that is not definitively proven for 60 years.

Source: Strandberg, J. (1920) "Regarding an unusual form of migrating erythema caused by tick bites." *Acta Derm Venereol (Stockh),* 1:422–427. Referenced in Weber/Burgdorfer (1993), chapter by Weber/Pfister.

1921
Arvid Afzelius, the Swedish dermatologist, emphasizes the link between tick bites and EM, describes an expanding rash with central clearing, and argues that the disease is self-limiting but can last up to 15 months.

Source: Afzelius, A. (1921) "Erythema chronicum migrans." *Acta Derm Venereol (Stockh),* 2:120–125. Referenced in Weber/Burgdorfer (1993), chapter by Weber/Pfister.

1921
An association between arthritis and ACA is reported for the first time.

Source: Jessner, M. (1921) "Zur Kenntnis der acrodermatitis chronica atrophicans." *Arch Dermatol Syph,* 134:478–487. Referenced in Weber/Burgdorfer (1993), chapter by Weber/Pfister.

1921
Severe cases of what is called *hyperplasia of cutaneous lymphoid tissue,* now known as a lymphocytoma, are reported.

Source: Kaufmann-Wolf, M. (1921) "Über gutartige lymphocytäre Neubildungen der Scrotalhaut des Kindes." *Arch Dermatol Syph,* 130:425–435. Referenced in Weber/Burgdorfer (1993), chapter by Hovmark et al.

1922
French physicians Charles Garin and Charles Bujadoux describe an association between EM and a painful neurologic condition called *meningoradiculoneuritis* (meningitis and severe radiating nerve pain) and paralysis of the deltoid muscle, now known to be a cranial nerve palsy. This is the first evidence of neurologic involvement associated with EM. The scientists incorrectly describe the disease as tick paralysis, but it is actually Bannwarth's syndrome.

Source: Garin, C. and C. Bujadoux. (1922) "Paralysie par les tiques." *J. Med Lyon,* 71:765–767. Referenced in Weber/Burgdorfer (1993), chapter by Weber/Pfister.

1923
The term *lymphocytoma* is used for the first time, based on a suggestion from M. Kaufmann-Wolf.

Source: Weber/Burgdorfer (1993), chapter by Hovmark et al.

1923
The first case of a patient with more than one EM rash is described.

Source: Lipschütz, B. (1923) "Weiterer Beitrag zur Kenntnis des erythema chronica migrans." *Arch Dermatol Syph,* 143:365–374. Referenced in Weber/Burgdorfer (1993), chapter by Weber/Pfister.

1924
Researchers M. Jessner and A. Loewenstamm report that 14% of ACA patients in one study had arthritic changes in their joints, especially joint thickening. Another study demonstrates that 59% of the ACA patients had arthritic bone atrophy in the areas of skin atrophy.

Source: Jessner, M. and A. Loewenstamm. (1924) "Bericht über 66 Fälle von Acrodermatitis chronica atrophicans." *Dermatol Wochenschr,* 79:1169–1177. Referenced in Weber/Burgdorfer (1993), chapter by Weber/Pfister.

1925
Researchers report patients having arthralgias prior to developing ACA and observe the similarities between that disease and syphilis, which is another spirochetal infection. They speculate that the disease is caused by a virus and is spread through the blood vessels and lymphatic system.

Source: Ehrmann, S. and F. Falkstein. (1925) "Über Dermatitis atrophicans und ihre pseudo-sklerodermatischen Formen." *Arch Dermatol Syph,* 149:142–175. Referenced in Weber/Burgdorfer (1993), chapter by Weber/Pfister.

1926
Atrophy of a hand joint is observed in an ACA patient.

Source: Piorkowski. (1926) "Zwei Fälle von acrodermatitis chronica atrophicans." *Zentralbl Haut,* 19:362. Referenced in Weber/Burgdorfer (1993), chapter by Weber/Pfister.

1929
Multiple lymphocytomas in association with ACA are reported.

Source: Mulzer, P. and E. Keining. (1929) "Über miliare lymphozytome der Haut." *Dermatol Wochenschr,* 88:293–301. Referenced in Weber/Burgdorfer (1993), chapter by Weber/Pfister.

1930
Sven Hellerström of the Karolinska Institute in Sweden links the EM rash and central nervous system conditions by reporting on a patient whose disease developed into

meningitis and encephalitis. Hellerström speculated that a spirochete is at work and may be found in the nervous system, not in the bloodstream, crediting this theory to the staining techniques of colleague Carl Lennhoff (see 1948 entry). This is the first time penicillin is known to have been used to cure a patient.

Source: Hellerström, S. (1930) "Erythema chronicum migrans Afzelii." *Acta Derm Venereol (Stockh),* 11:315–321. Referenced in Weber/Burgdorfer (1993), chapter by Weber/Pfister.

1931
ACA, which causes a sclerosis process, is distinguished from true scleroderma, which is a hardening and contracting of the body's connective tissue, because ACA sclerosis appears in areas of preexisting atrophy and does not occur with any other condition.

Source: Oppenheim, M. (1931) "Atrophien." *Handbuch der Haut-und Geschlechts Krankheiten.* J. Jadassohn Editor, Berlin: Springer; 8/2: 500–716. Referenced in Scrimenti (1995).

1931
Joint atrophy seen in ACA patients is described as involving a first stage of inflammation and a second stage of atrophy.

Source: Hövelborn, C. I. (1931) "Gelenkveränderungen bei acrodermatitis chronica atrophicans." *Arch Dermatol,* 164:349–356. Referenced in Weber/Burgdorfer (1993), chapter by Åsbrink et al.

1934
R. Stadelmann, a German physician, expands the thinking about EM disease to include inflammatory arthritis, skin rash, muscle pain, and fatigue in a description that highlights the multisystem nature of disease. A patient with the EM rash and simultaneous heart problems is also described.

Source: Stadelmann, R. (1934) *Ein Beitrag zum Krankheitsbild des erythema chronicum migrans Lipschütz.* Dissertation, University of Marburg. Referenced in Weber/Burgdorfer (1993), chapter by Weber/Pfister.

1935
United States physicians report hardening of connective tissue as an integral part of ACA. They also identify other associated symptoms, including extreme hair loss; skin pigmentation abnormalities; increased sensitivity to sensory stimulation; arthritic atrophy of the fingers, hands, and toes; and disruption of the mucous membrane of the nose, tongue, throat, and vagina.

Source: Stadelmann, R. (1934) *Ein Beitrag zum Krankheitsbild des erythema chronicum migrans Lipschütz.* Dissertation, University of Marburg. Referenced in Weber/Burgdorfer (1993), chapter by Weber/Pfister.

1936
H. Askani, a German dermatologist, presents data suggesting that tick bites are the probable cause of the EM rash. He hypothesizes that the infection is found in the tick's salivary glands and is either a toxin or a living pathogen.

Source: Askani, H. (1936) "Zur Ätiologie des Erythema chronicum migrans." *Dermatol Wochenschr,* 102:125–131. Referenced in Weber/Burgdorfer (1993), chapter by Burgdorfer.

1941
Alfred Bannwarth, a German neurologist, describes a cluster of chronic or recurring neurologic symptoms that he called *chronic lymphocytic meningitis* and that is now recognized as the classic neurological LD triad of severe nerve pain, cranial nerve palsy, and meningitis, which is known as Bannwarth's syndrome.

Source: Bannwarth, A. (1941) "Chronische lymphocytäre meningitis, entzündliche polyneuritis und rheumatismus." *Arch/Psychiatr Nervenkr,* 113:284–376. Referenced in Weber/Burgdorfer (1993), chapter by Weber/Pfister.

1942
At a medical meeting, R. Kahle, a German dermatologist, reports that the blood serum from six of his seven ACA patients is positive for syphilis, suggesting that there may be a spirochetal origin to the disease.

Source: Kahle, R. H. (1942) *Pallida-Reaktion bei peripheren Durchblutungsstörungen der Haut, insbesondere bei Acrodermatitis atrophicans.* Dissertation, University of Halle. Referenced in Weber/Burgdorfer (1993), chapter by Burgdorfer.

1943
Swedish researcher Bo Bäfverstedt describes the appearance and location of benign skin swellings filled with white blood cells that he calls *lymphadenosis benigna cutis* (still referred to as "LABC") in 150 patients, including several with EM or ACA. He describes the disease as having a localized and disseminated form.

Source: Bäfverstedt, B. (1943) "Über Lymphadenosis benigna cutis." *Acta Derm Venereol (Stockh) (Suppl),* 24:1–202. Referenced in Weber/Burgdorfer (1993), chapter by Weber/Pfister.

1945
Mayo Clinic researchers present additional cases of ACA. Although some of the patients had arthralgia, arthritis, and heart problems prior to ACA, the researchers conclude they are unrelated to the skin disorder.

Source: Montgomery, H. and R. R. Sullivan. (1945) "Acrodermatitis atrophicans chronica." *Arch Dermatol,* 51:32–47. Referenced in Weber/Burgdorfer (1993), chapter by Weber/Pfister.

1946
Penicillin is demonstrated to be an effective treatment for *ACA Herxheimer.*

Source: Svartz, N. (1946) "Penicillinbehandling vid dermatitis atrophicans Herxheimer." *Nord Med,* 32:2783. Referenced in Weber/Burgdorfer (1993), chapter by Weber/Pfister.

1948
Carl Lennhoff, of Stockholm's Karolinska Institute (where research on Lyme disease and related skin conditions continues today), describes a staining technique that demonstrated spirochetes in 27 different diseases, including EM skin biopsies. Lennhoff's work intensified efforts to prove that a spirochete was the causative agent of EM.

Source: Lennhoff, C. (1948) "Spirochetes in aetiologically obscure diseases." *Acta Derm Venereol (Stockh),* 28:295–324. Referenced in Weber/Burgdorfer (1993), chapter by Burgdorfer.

1949
R. Schuppli of the Dermatologic Clinic of the State Hospital in Basel, Switzerland, goes to Lennhoff's Stockholm lab to study his staining technique. Although he was then able to stain spirochetes from the tissue of the EM rash, Schuppli was dissatisfied with the technique and criticized it publicly. That closed the door on a technique that might have provided answers about the cause of the rash.

Source: Weber/Burgdorfer (1993), chapter by Burgdorfer.

1949
At the Southern Medical Association in Ohio, Sven Hellerström describes tick bites that result in both the rash and neurologic problems (which he referred to as ECM Afzelius with meningitis). Hellerström reports on the successful use of antibiotics and discusses the possibility that an *Ixodes* tick could transmit spirochetes, a theory that is considered preposterous. Although the presentation was made in the United States, there are still no cases of EM rash known to be acquired in this country.

Source: Hellerström, S. (1950) "Erythema chronicum migrans Afzelius with meningitis." *Southern Medical Journal,* 43:330–334. Referenced Weber/Burgdorfer (1993), chapter by Burgdorfer.

1949
Niels Thyresson of Stockholm notes that penicillin appears to be an effective treatment for ACA, reporting that of 57 patients studied, 7 were cured, 28 improved, and 5 showed slight improvement. The fibrous nodules disappeared and sensory deficits improved.

Source: Thyresson, N. (1949) "The penicillin treatment of acrodermatitis chronica atrophicans (Herxheimer)." *Acta Derm Venereol (Stockh),* 29:572–621. Referenced in Weber/Burgdorfer (1993), chapter by Weber/Pfister.

1949

German researcher George Schaltenbrand reports on eight patients with subacute meningitis, one of whom also had arthritis. After the death of one of the patients, a 36-year-old woman, an autopsy was performed and doctors found widespread vasculitis, which is serious inflammation of some of the blood vessels in the brain.

Source: Schaltenbrand, G. (1949) "Chronische aseptische meningitis." *Nervenarzt,* 20:433–442. Referenced in Weber/Burgdorfer (1993), chapter by Weber/Pfister.

1950

Good results for treating lymphocytoma patients with penicillin are reported.

Source: Bianchi, G. E. (1950) "Die penicillinbehandlung der Lymphocytome." *Dermatologica,* 100:270–273. Referenced in Weber/Burgdorfer (1993), chapter by Weber/Pfister.

1951

The term *Zeckenbiss-lymphocytoma* (tick bite lymphocytoma) is proposed for the lymphocytomas caused by tick bites.

Source: Jordan, P. and J. Holtschmidt. (1951) "Traumatisches Zeckenbiss-Lymphocytom und erythema chronicum migrans." *Hautarzt,* 2:397–401. Referenced in Weber/Burgdorfer (1993), chapter by Hovmark et al.

1951

Low-dose intravenous penicillin is reported to be effective in treating EM patients, including one with meningitis. The recovery time averaged about 2 weeks.

Source: Hollström. "Successful treatment of EM Afzelius." *Acta Derm Venereol (Stockh),* 31:235–289. Referenced in Weber/Burgdorfer (1993), chapter by Weber/Pfister.

1951

An exacerbation of symptoms, including a rise in temperature, after treatment begins is noted in an ACA patient and is similar to what has been described in syphilis. This fever of unknown origin is now called LD Jarisch-Herxheimer reaction.

Source: Götz, H. and E. Ludwig. (1951) "Die Behandlung der Akrodermatitis chronica atrophicans Herxheimer mit penicillin." *Hautarzt,* 2:6–14. Referenced in Weber/Burgdorfer (1993), chapter by Weber/Pfister.

1952

For months prior to developing ACA, a patient is reported to have arthritis in several joints, as well as swollen glands, psychological changes, heart problems, muscle inflammation, bone loss in the joints of her fingers, and impaired movement of her shoulders and elbows.

Source: Gans, O. and E. Landes. (1952) "Akrodermatitis atrophicans arthopatica." *Hautarzt,* 3:151–155. Referenced in Weber/Burgdorfer (1993), chapter by Weber/Pfister.

1952
European researcher T. Grüneberg postulates that the positive reaction of ACA Herxheimer patients to the syphilis test suggests the involvement of a spirochete.

Source: Grüneberg, T. (1952) "Zur Frage der Ätiologie der acrodermatitis chronica atrophicans." *Dermatol Wochenschr,* 126:1041–1046. Referenced in Weber/Burgdorfer (1993), chapter by Burgdorfer.

1952
German researcher Walter Hauser identified inflammatory changes in the bone marrow of 25 ACA patients.

Source: Hauser, W. (1952) "Sternalmarkbefunde und ihre Beziehungen zur Blutsenkungsgeschwindigkeit bei acrodermatitis chronica atrophicans." *Arch Dermatol Syph,* 195:164–170. Referenced in Weber/Burgdorfer (1993), chapter by Weber/Pfister.

1954
Swiss physician J. Paschoud related a case of tick bite that resulted in a lymphocytoma and was followed by Bannwarth's syndrome.

Source: Paschoud, J. M. (1954) "Lymphocytom nach Zeckenbiss." *Dermatologica,* 108:435–437. Referenced in Weber/Burgdorfer (1993), chapter by Weber/Pfister.

1955
The infectious nature of ACA is demonstrated when German dermatologist Hans Götz successfully transmits the disease by transplanting ACA-infected skin to healthy volunteers, including himself. Resulting symptoms include oversensitivity to sensory stimulation and joint problems. Götz notes that the disease looks like EM disease when it starts and speculates that it is disseminated throughout the body, much like syphilis.

Source: Götz, H. (1955) "Die Acrodermatitis chronica atrophicans Herxheimer als Infektionskrankheit." *Hautarzt,* 5:491–504, 6:249–252. Referenced in Weber/Burgdorfer (1993), chapter by Weber/Pfister.

1955
Blood from ACA patients is inoculated into mice that are then found to produce antibodies against spirochetes. However, other researchers are unable to reproduce this study so the claim that the skin condition is caused by a spirochete is rejected.

Source: Lohel, H. (1955) "Tierexperimentelle Untersuchungen zur Ätiologie der acrodermatitis chronica atrophicans Herxheimer." *Klin Wochenschr,* 33:185–186. Referenced in Weber/Burgdorfer (1993), chapter by Burgdorfer.

1955
The first proof that EM is an infectious disease becomes available after researchers transplant pieces of EM tissue samples from infected patients onto the skin of healthy volunteers. These volunteers develop the characteristic rash within 1 to 3 weeks.

Researchers are also able to infect an additional person by using the skin of the infected volunteer. Treatment with penicillin resolves the rash successfully.

Source: Binder, E., R. Doepfmer and O. Hornstein. (1955) "Experimentelle Übertragung des erythema chronicum migrans von Mensch zu Mensch." *Hautarzt,* 6:494–496. Referenced in Weber/Burgdorfer (1993), chapter by Weber/Pfister.

1955
German researcher Walter Hauser explores the relationship among ACA, EM, and lymphocytomas and notes that some patients develop bone-deforming arthritis. He concludes that these diseases are somehow related and that ACA is a systemic disease. He also suggests a correlation between ACA cases and the distribution of sheep *(Ixodes)* ticks.

Source: Hauser, W. (1955) "Zur Kenntnis der akrodermatitis chronica atrophicans." *Arch Dermatol Syph,* 199:350–393. Referenced in Weber/Burgdorfer (1993), chapter by Weber/Pfister.

1956
A detailed description of EM is provided in a medical textbook in the United States, making information about this disease accessible to the United States medical community for the first time. All reported cases are still being identified only in Europe.

Source: Sutton (1956)

1957
Lymphocytoma contents from a patient who was developing the disease are injected into a healthy volunteer, proving that the disease is infectious. The Swiss researcher found that if he injected the pathogen deep into the skin, the patient developed lymphocytomas, whereas a shallow injection resulted in an EM.

Source: Pashoud, J. M. (1957) "Die lymphadenosis benigna cutis als übertragbare infektionskrankheit." *Hautarzt,* 8:197–211, 9:153–165, 263–269, 311–315. Referenced in Weber/Burgdorfer (1993), chapter by Hovmark et al.

1958
EM patients given low dose intravenous penicillin are reported to be having relapses. Most patients were women with an average age of 43 years.

Source: Hollström, E. (1958) "Penicillin treatment of erythema chronicum migrans Afzelius." *Acta Derm Venerol (Stockh),* 38:285–289. Referenced in Weber/Burgdorfer (1993), chapter by Weber/Pfister.

1959
After reviewing 800 cases of ACA reported in the academic literature, researchers report that approximately 5 years after the skin rash, fibroid nodules and skin degeneration and discoloration appear on the arms, and atrophy of the skin occurs, primarily on the legs.

Source: Donnerman, C. and H. J. Heite. (1959) "Beitrag zur Symptomatologie der akrodermatitis chronica atrophicans (Pick-Herxheimer)." *Arch Klin Exp Dermatol,* 208:516–527. Referenced in Weber/Burgdorfer (1993), chapter by Weber/Pfister.

1960
A correlation is made between EM rash and the peak season for *Ixodes* tick bites. In the same research, it is noted that a small percentage of patients develop an ACA-like condition after treatment.

Source: Sedlacek, V. (1960) "Erythema chronicum migrans-poznamky ke klinice, etiopatogenezi a zarazeni." *Cs Dermatol,* 35:386–399. Referenced in Weber/Burgdorfer (1993), chapter by Weber/Pfister.

1962
German researcher adds to the expanding knowledge of Bannwarth's syndrome by presenting cases in which tick bites are followed by redness in the surrounding skin and the development of neurological symptoms. Patients improve after using tetracycline.

Source: Schaltenbrand, G. (1962) "Radikulomyelomenigitis nach Zeckenbiß." *Münch Med Wochenschr,* 104:829–834. Referenced in Weber/Burgdorfer (1993), chapter by Weber/Pfister.

1962
French physicians add to a growing debate about the pathogenic source of disease by showing that 6 of 7 EM patients tested positive to various forms of rickettsial agents.

Source: Dégos, R., R. Tourraine and J. Arouete. (1962) "L'erythema chronicum migrans." *Syph,* 89:247–260. Referenced in Weber/Burgdorfer (1993), chapter by Burgdorfer.

1963
An ACA patient with positive results from a rheumatoid test is described.

Source: Brehm, G. (1963) "Symptomatische Makro- und Kryoglobulinämie bei akrodermatitis chronica atrophicans." *Hautarzt,* 14:75–79. Referenced in Weber/Burgdorfer (1993), chapter by Weber/Pfister.

1965
Four patients whose tick bites were followed by EM and subsequent central nervous system involvement are described by a researcher.

Source: Bammer, H. and K. Schenk. (1965) "Menigo-myelo-radikulitis nach Zeckenbiß mit erythem." *Dtsch Z Nervenheilkd,* 187:25–34. Referenced in Weber/Burgdorfer (1993), chapter by Weber/Pfister.

1965
For the first time, a researcher suggests that the three skin conditions of ACA, EM, and lymphocytoma represent one disease entity, consolidating the various symptoms

we now know as Lyme disease. The researcher also emphasized the role of the tick as the vector. This theory was proved true when the causative agent was identified, almost 20 years later.

Source: Hauser, W. (1965) "Wahrscheinliche Infektionskrankheiten der Haut." *Handbuch Haut-u. Geschlechtskrankheiten,* Vol. IV, Part 1A, pp 556–629. Referenced in Weber/Burgdorfer (1993), chapter by Weber/Pfister.

1966
A report on ten children with Bell's palsy linked to a condition then being called subacute chronic lymphocytic meningitis, or Bannwarth's syndrome, is published.

Source: Keuth, U. and U. Mennicken. (1966) "Zur Kenntnis der peripheren Facialisparese mit subakut-chronisch lymphozytärer Meningitis (Bannwarth)." *Z. Kinderheilkd,* 97: 45–56. Referenced in Weber/Burgdorfer (1993), chapter by Weber/Pfister.

1966
After feeding *Ixodes* ticks at the larval, nymphal, and adult stages, a researcher reports no evidence of detectable spirochetes and finds that none of the nymphal ticks are able to acquire infection and transmit disease to another host. This is seen as proof that spirochetes were not involved and suggests that a mosquito can transmit the EM rash.

Source: Hard, S. (1966) "Erythema chronicum migrans (*Afzelli*) associated with mosquito bite." *Acta Derm Venereol (Stockh),* 46:473–476. Referenced in Weber/Burgdorfer (1993), chapter by Burgdorfer.

1966
German researcher Hanns Christian Hopf theorizes an association between peripheral neuropathy and ACA. Hopf also reports that neurologic symptoms begin an average of 4 months after the EM rash and that most patients have normal findings on a spinal tap. Other associated symptoms include arthritis (in 26% of the patients studied), cardiac involvement (19%), and neurologic problems (11%). Penicillin is reported to help the majority of patients.

Source: Hopf, H. C. and B. Stroux. (1968) "Die geographische Verteilung der Akrodermatitis chronica atrophicans (Herxheimer) in der Umgebung von Würzburg." *Z. Hautkr,* 43:41–48. Referenced in Weber/Burgdorfer (1993), chapter by Weber/Pfister.

1967
For the first time, a clear distinction is made between EM-related nervous system disorders and tick-borne encephalitis. Common signs of the EM neurologic condition are described as inflammation of the nerves, inflammation of the spinal cord, and cranial nerve palsy. G. Schaltenbrand asserts that the *Ixodes* tick is probably the primary vector, but that about 10% of the patients get the disease from the bite of a horsefly.

Source: Schaltenbrand, G. (1967) "Durch Artropoden übertragene Erkrankung der Haut und des Nervensystems." *Verhandl Dtsch Ges Inn Med,* 72:975–1006. Referenced in Weber/Burgdorfer (1993), chapter by Weber/Pfister.

1968
The first clustering of more than 300 ACA cases is identified and described in Germany.

Source: Hopf, H.C. and B. Stroux. (1968) "Die geographische Verteilung der akrodermatitis chronica atrophicans (Herxheimer) in der Umgebung von Würzburg." *Z. Hautkr,* 43:41–48. Referenced in Weber/Burgdorfer (1993), chapter by Weber/Pfister.

1968
Another report is published on the multitude of neurological symptoms that may follow tick or insect bites and the EM rash. Specifically noted are stiff neck, headache, excessive sleepiness, mood swings, loss of contact with reality, paranoia, and hallucinations.

Source: Erbslöh, F. and K. Kohlmeyer. (1968) "Über polytope Erkrankungen des peripheren Nervensystems bei lyphocytärer Meningitis." *Fortschr Neurol Psychiatr,* 36:321–342. Referenced in Weber/Burgdorfer (1993), chapter by Weber/Pfister.

1970
Cases of Bannwarth's cranial nerve paralysis are reported, as well as the fact that two of the patients attempted suicide because of the pain.

Source: Wolf, G. (1970) "Über die chronische lymphocytäre Meningitis unter dem Bilde der Polyneuritis (Bannwarth)." *Fortschr Neurol Psychiatr,* 38:221–234. Referenced in Weber/Burgdorfer (1993), chapter by Weber/Pfister.

1970
Rudolph Scrimenti, associate clinical professor in the Department of Dermatology at the Medical College of Wisconsin, and an expert on LD skin infection, reports the first instance of an EM rash known to be acquired in the United States. The patient was a physician who had been grouse hunting in Wisconsin and had removed small, engorged ticks from his body. In his report, Scrimenti describes the accompanying neurologic and arthritic symptoms and discusses the use of penicillin as treatment.

Source: Scrimenti, R. (1970)

1971
Techniques and medium for culturing *Borrelia* are established. Richard Kelly modified a syphilis medium formula by adding the ingredient N-acetyl glucosamine, a building block of the tick's hard shell.

Source: Kelly, R. (1971) "Cultivation of *Borrelia hermsii.*" *Science,* 173:443–444. Referenced in Weber/Burgdorfer (1993), chapter by Burgdorfer.

1972

ACA patients treated with penicillin are reported to have a relapse of peripheral neuropathy.

Source: Kaiser, M. (1972) *Neurologische Komplikationen bei Akrodermatitis chronica atrophicans (Herxheimer) und ihre Beeinflussung durch die Penicillintherapie.* Dissertation, University of Göttingen. Referenced in Weber/Burgdorfer (1993), chapter by Weber/Pfister.

1974

Klaus Weber, a German physician, presents a case of EM that develops into meningitis, despite treatment with oral penicillin. Treatment with high-dose intravenous penicillin ultimately results in a cure.

Source: Weber/Burgdorfer (1993), chapter by Weber.

1974

A team of physicians finds evidence that nerve conduction tests may be normal in patients with Bannwarth's syndrome but they may nonetheless have damage to the nerve itself.

Source: J. Rohmer, M. Collard, M. Jesel, J. Walter, G. Coquillat and J. Glass. (1974) "Les meningoradiculites: données clinques, electromyographiques et etiologiques à propos de 36 observations. Limites nosologiques." *Rev Neurol,* 130:415–431. Referenced in Weber/Burgdorfer (1993), chapter by Weber/Pfister.

1975

ACA patients are found to have bone damage and radiating nerve pain reinforcing the multisystemic nature of the disease.

Source: Hopf, H. (1975) "Peripheral neuropathy in acrodermatitis chronica atrophicans (Herxheimer)." *Journal of Neurosurg Psychiat,* 38:452–458. Referenced in "Acrodermatitis chronica atrophicans." E. Åsbrink, A. Hovmark and K. Weber. *Aspects of Lyme Borreliosis.* Edited by K. Weber and W. Burgdorfer. Heidelberg, Germany: Springer-Verlag, 1993. pp. 201–204.

1975

Polly Murray and her family, residents of Lyme, Connecticut, have been sick since the 1960s with a variety of ailments, including rashes, swollen knees, and sore throats. In her search for answers, Polly contacts Yale University. About the same time and in the same area, Judy Mensch notices a cluster of juvenile rheumatoid arthritis cases. To report this unusual observation, she contacts the Connecticut State Department of Health, which also turns to Yale University for help.

Allen Steere, a former Centers for Disease Control epidemiologist working at Yale, is asked to lead an investigation into this novel outbreak.

1975
Rudolph Scrimenti hears about the Connecticut investigation and sends scientific information about the EM rash to the Yale team.

1976
Researchers at the Naval Submarine Medical Center describe a cluster of EM cases in southeastern Connecticut. Although none of these patients remembers being bitten, researchers speculate an insect-borne infectious disease is involved.

Source: Mast and Burrows (1976)

1977
Allen Steere and his colleagues at Yale University publish the results of a study open only to patients with recurrent attacks of joint swelling, of 39 children and 12 adults in three Connecticut towns, of whom 25% had an EM rash. After reviewing their symptoms, the authors claim to have identified a new clinical entity and name it Lyme arthritis, after one of the Connecticut towns in which it was found. Joint swelling was declared a major characteristic of this disease, hardly surprising since it was a necessary qualification for enrollment. Four patients had been treated with short-term antibiotics prior to the investigation yet developed arthritis, leading the researchers to conclude that antibiotics didn't work. Patients are treated with aspirin.

Source: Steere et al. (1977b)

1977
In a follow-up study, Steere and his colleagues report on more patients with the clinical entity they call LD. In addition to arthritic problems, the majority of these patients had some type of neurologic involvement, severe fatigue, and EM rashes. The authors question the value of antibiotic treatment, pointing out that 25% of those enrolled in their study had previously used short-term antibiotics yet continued to have symptoms. The researchers decide against treating the patients with antibiotics, despite citing numerous European studies attesting to their effectiveness and just one claiming they are of no value. Steroids and aspirin are used as treatment, although the researchers express some reservations about that therapeutic approach.

Source: Steere et al. (1977a)

1978
In a study of *Ixodes scapularis* ticks from the southeastern Connecticut area, researchers report that tests for rickettsiae, viruses, and other pathogens are all negative.

Source: Wallis et al. (1978)

1979
Rickettsia-like organisms are found in the immune system cells of two EM patients. Retrospectively, this may be the first evidence of coinfection with another tick-transmitted disease.

Source: Sandbank, M. and E. F. Feuerman. (1979) "Ultrastructural observation of rickettsia-like bodies in erythema chronicum migrans." *J Cutan Pathol,* 6:253–264. Referenced in Weber/Burgdorfer (1993), chapter by Burgdorfer.

1979
Recognizing the broadening spectrum of the disease, Allen Steere and Steven Malawista change its name from Lyme arthritis to Lyme disease.

Source: Steere and Malawista (1979)

1979
Researchers claim to have found a new species of tick in the northeastern part of the United States and name it *Ixodes dammini.* However, many entomologists believe the ticks are nothing more than another sample of *Ixodes scapularis,* the blacklegged tick. That view is affirmed in 1993 and the original name is officially reinstated.

Source: Spielman et al. (1979)

1981
German physician Klaus Weber has the blood of thirteen EM patients tested for rickettsiae in two laboratories, the Rocky Mountain Laboratories of the National Institutes of Health and the Institute of Virology in Bratislava, Czechoslovakia. All tests are negative. He postulates that persisting infection is involved and recommends antibiotic therapy.

Source: Weber, K. (1981) "Serological study with rickettsial antigens on erythema chronicum migrans." *Dermatologica,* 163:460–467. Referenced in Weber/Burgdorfer (1993), chapter by Burgdorfer.

1981
Breakthrough! A new spirochete is found in *Ixodes scapularis* ticks, collected by Edward Bosler of the New York State Department of Health in the ongoing project aimed at identifying both the pathogen that causes LD and the vector that transmits that pathogen. Credit goes to Willy Burgdorfer, an entomologist at the Rocky Mountain Laboratories (and later a founding board member of the Lyme Disease Foundation [LDF]).

1982
Researchers at the Rocky Mountain Laboratories modify Kelly's *Borrelia* medium so that it can be used to culture *Borrelia hermsii,* which causes relapsing fever. The new medium is called Stoenner-Kelly medium.

Source: Stoenner, H. G., T. Dodd and C. Larsen. (1982) "Antigenic variation of *Borrelia hermsii.*" *J Exp Med,* 156:1297–1311. Referenced in Weber/Burgdorfer (1993), chapter by Burgdorfer.

1982
Rudy Ackermann, a German neurologist, finds spirochetes in *Ixodes ricinus* ticks in two places in Germany. Ackermann is credited with successfully treating EM and other manifestations of disease with penicillin and tetracycline. He speculates that the relapsing nature of the disease suggested a spirochetal etiology.

Source: Ackermann, R., J. Kabatzki, H. P. Boisten, A.C. Steere, R.L. Grodzicki, S. Hartung and U. Runne. (1984) "Spirochäten-Ätiologie der erythema-chronicum-migrans-Krankheit." *Dtsch Med Wochenschr,* 109:92–97. Referenced in Weber/Burgdorfer (1993), chapter by Weber/Pfister.

1982
Willy Burgdorfer publishes his paper on the discovery of the causative agent of Lyme disease and earns the right to have his name placed on the spirochete. *Borrelia burgdorferi* is now the official designation.

Source: Burgdorfer et al. (1982)

1982
A long-term study of people with EM shows that 32% develop arthritis and joint pain, 23% have neurologic problems, and 10% have cardiac involvement.

Source: Felgenhauer, K. (1982) "Differentiation to the humoral response in inflammatory diseases of the CNS." *J. Neurol,* 228:223–227. Referenced in Weber/Burgdorfer (1993), chapter by Weber/Pfister.

1982
The Centers for Disease Control establishes the first federal reporting for Lyme disease, which is voluntary. Individual states are allowed to develop their own disease definitions.

Source: "Lyme Disease—United States, 1995." (1996)

1982
Australia reports its first case of what we now call Lyme disease, which involves a 21-year-old laborer.

1983
U.S. dermatologist Bernard Berger and his colleagues find spirochetes in EM lesions.

Source: Berger et al. (1983)

1983
Paul Duray, an American pathologist, perfects a tissue staining technique and is the first person to see *Bb* in tissue other than skin.

1983
Willy Burgdorfer finds *Bb* in Swiss *Ixodes* (sheep) ticks that is identical to the *Bb*

found in *Ixodes scapularis* (black-legged) ticks on Shelter Island, New York. This proves the European and U.S. diseases are the same.

Source: Weber/Burgdorfer (1993), chapter by Burgdorfer.

1983
Bb is isolated from patients with Lyme disease in Connecticut.

Source: Steere et al. (1983)

1983
Rudy Ackermann finds antibodies to *Bb* in patients with EM rashes and meningoradiculoneuritis (Bannwarth's).

Source: Ackermann, R. (1983) "Erythema chronicum migrans und durch Zecken übertragene meningopolyneuritis (Garin-Bujadoux-Bannwarth): Borrelien Infektionen?" *Dtsch Med Wochenschr,* 108:577–580. Referenced in Weber/Burgdorfer (1993), chapter by Weber/Pfister.

1983
Jorge Benach, a researcher at the New York State Department of Health, and his colleagues find *Bb* spirochetes in the blood of two LD patients.

Source: Benach (1983)

1983
Willy Burgdorfer predicts that Lyme disease could become the number-one tick-borne disease in the United States and is ultimately proven correct.

Source: W. Burgdorfer (1984)

1983
Positive blood tests for *Bb* are found in patients with erythema migrans, acrodermatitis chronica atrophicans, and lymphocytoma.

Source: Weber, K., G. Schierz, B. Wilske, B. Preac-Mursic, W. Burgdorfer, and A. Barbour. (1983/84) "Antibodies against *Ixodes dammini* and *Ixodes ricinus* spirochetes in tick-borne disorders." 11th ADF meeting, Kiel, Nov. 1983. *Arch Dermatol Res,* (1984) 276.260.

1984
Alan Barbour, working at Rocky Mountain Laboratories, cultures the first LD spirochetes from ticks from Shelter Island, New York, and names the isolate B–31. The newly developed medium is called Barbour-Stoenner-Kelly I.

Source: Barbour (1984)

1984
Swedish dermatologists at the Karolinska Institute culture *Bb* spirochetes from EM and ACA lesions, thus confirming that they are part of the enlarging spectrum known as Lyme borreliosis.

Source: Åsbrink, E., A. Hovmark and B. Henderstedt. (1984) "The spirochetal etiology of acrodermatitis chronica atrophicans Herxheimer." *Acta Derm Venereol, (Stockh)* 64:506–512. Referenced in Weber/Burgdorfer (1993), chapter by Weber/ Pfister.

1984
Bb is isolated from the spinal fluid of a meningoradiculoneuritis patient by German physicians.

Source: Pfister, H. W., K. M. Einhäupl, V. Preac-Mursic, B. Wilske and G. Schierz. (1984) "The spirochetal etiology of lymphocytic meningoradiculitis of Bannwarth (Bannwarth's syndrome)." *J. Neurol,* 231:141–144. Referenced in Weber/Burgdorfer (1993), chapter by Weber/Pfister.

1984
Bb organisms are isolated from the blood of EM patients by a German neurologist, proving that EM is part of Lyme disease. The physician is able to demonstrate that EM patients test negative to *Bb* antibody tests.

Source: Ackermann, R., H. P. Boisten, J. Kabatzki, U. Runne, K. Krüger and W. P. Herrmann. (1984) "Serumantikörper gegen *Ixcodes-ricinus* Spirochäte bei acrodermatitis chronica atrophicans (Herxheimer)." *Dtsch Med Wochenschr,* 109: 6–10. Referenced in Weber/Burgdorfer (1993), chapter by Åsbrink et al.

1984
The worsening of symptoms at the beginning of treatment is again described, this time in patients treated with oral penicillin and minocycline.

Source: Weber, K. (1984) "Jarisch-Herxheimer-Reaktion bei erythema migrans-Krankheit." *Hautarzt,* 35:588–590. Referenced in Weber/Burgdorfer (1993), chapter by Burgdorfer.

1984
Researchers discover the Lyme disease pathogen in lone star ticks in New Jersey.

Source: Schulze et al. (1984)

1984
Russell Johnson, a microbiologist from the University of Minnesota, sequences the DNA of this spirochete and declares it a new *Borrelia* species. He proposes the name *Borrelia burgdorferi* sp. nov. "Sp. nov." is used until a taxonomy group declares a new name to be official.

Source: Johnson et al. (1984)

1985
Willy Burgdorfer finds *Bb* in Western black-legged ticks, proving the spirochete is spread across the country.

Source: Weber/Burgdorfer (1993), chapter by Burgdorfer.

1985
Centres for Disease Control of British Columbia begin research into LD, collaborating with Rocky Mountain Laboratories.

1985
At the Second International Lyme Borreliosis conference, Paul Duray declares that *Bb* disseminates early and infects more body organs than previously recognized. This theory conflicts with the prevailing viewpoint that the infection takes a long time to disseminate, but it is ultimately proven correct.

1985
Researchers publish the first proof of maternal-fetal transmission of *Bb*. The baby died shortly after birth and *Bb* was demonstrated in the baby's spleen, kidney, and bone marrow.

Source: Schlesinger et al. (1985)

1985
Karen Vanderhoof-Forschner, author of this book, is bitten by something and develops a circular rash. She is about 50 miles away from Yale University, and no one can identify the rash. Karen develops many problems, including severe headaches, profound fatigue, serious joint swelling, disturbances in taste and smell, severe radiating leg pains, and fevers. She is plagued by pre-term labor and delivers a son in July. Because of profound joint swelling she consults another doctor who says that she may have LD. Her son starts developing serious medical problems (including eye tremors, vomiting, nerve conduction problems, and a rash) but she is told that LD cannot be passed to a developing fetus.

Her experiences begin the cascade of events that eventually lead to the creation of the Lyme Disease Foundation.

1986
Three more cases of Lyme disease are reported in Australia.

1986
Hungarian physician Bela Bózsik and his colleagues report on neurologically infected people with *Bb* who are cured by antibiotics. They also document that a negative Lyme test cannot differentiate between Lyme borreliosis and tick-borne encephalomyelitis. This is discouraging, since both diseases are spread by *Ixodes* (sheep) ticks and have similar symptoms but are treated very differently.

Resentment begins to emerge in the international research community because this disease has been recognized and researched in Europe for more than a century but is now being given a new name by a U.S. researcher who is credited with discovering it. Europeans use the term borreliosis instead of Lyme disease. Bela Bózsik proposes a compromise name—"Lyme borreliosis"—to the international community.

1986
The first case report of ACA acquired in the United Staes is described by California rheumatologist Paul Lavoie and his colleagues.

Source: Lavoie (1986)

1986
Vera Preac-Mursic, a German researcher from the Max von Pettenkofer Institute, modifies the Kelly *Borrelia* culture medium in order to grow European strains of *Bb.* This new medium is called Preac-Mursic (MKP) medium. This enables the European community to conduct more spirochetal research.

Source: Preac-Mursic (1986)

1986
Bb is isolated from a patient with borrelia lymphocytoma, proving an association that had only been suspected in the past.

Source: Hovmark, A., E. Åsbrink and I. Olsson. (1986) "The spirochetal etiology of lymphadenosis benigna cutis solitaria." *Acta Derm Venereol (Stockh),* 66:479–484. Referenced in Weber/Burgdorfer (1993), chapter by Burgdorfer.

1986
Bb is cultured from the blood of a patient with an EM rash.

Source: Neubert, U., H. E. Krampitz and H. Engl. (1986) "Microbiological findings in erythema chronicum migrans, acrodermatitis chronica atrophicans und lymphadenosis cutis benigna." *Zentralbl Bakteriol Hyg* (A), 263:237–252. Referenced in Weber/Burgdorfer (1993), chapter by Weber/Pfister.

1986
A period of activism begins. Vicki Korman (a founding board member of the LDF) and Carol Gabriel, both Lyme disease patients with serious multisystem problems in New Jersey, and novice activists, join forces. Terry Schulze, an entomologist with the state Department of Health, gets involved, and the team tapes an education program about LD that is distributed to local hospitals. Later, they entice *20/20* to produce a television segment on LD.

1986
Gloria Wenk of Dutchess County, New York, organizes a petition drive to express concern about tick problems in the area. Although county health officials tell her no problem exists, she secures 1,800 signatures from local residents. By mid-1986, Wenk is increasing her community tick awareness level and is widely known as the "Tick Lady."

1988
The Lyme Borreliosis Foundation, the first organization in the world dedicated exclusively to Lyme disease, is incorporated. The foundation was proposed by three

researchers who felt that accurate scientific information was not being disseminated, partly due to excessive infighting among scientists. The foundation's goal is to increase awareness about Lyme disease through education and to promote research and international collaborations.

The foundation sponsors its first annual international scientific conference, and the television program *20/20* produces a segment about the organization. After the show airs, the foundation is flooded by telephone calls and letters from around the world. More international media coverage quickly follows. The foundation succeeds in securing the first federal funds dedicated to LD and convinces Congress to pass a resolution declaring national Lyme Disease Awareness Week. It also establishes the first network of support groups and fosters the development of Lyme organizations around the world.

1988
With many children becoming ill in her Pennsylvania neighborhood, Amy Jones contacts the LDF and launches her own local campaign against LD. Her successful lobbying and fund-raising efforts do much to call attention to the problem of LD in Pennsylvania.

1988
British Columbia Centres for Disease Control discovers 13 positive LD cases. Three could be proved as being acquired in British Columbia.

Source: Banerjee (1993a)

1989
Diane Kindree, R.N., and her physician father, L. Kindree, launch the Vancouver, B.C., Lyme Borreliosis Society (LBS) at the urging of the LDF and with the support of her family and a team of local activists. The organization is dedicated to collaborating with scientists, educating the medical community/general public, and providing support to patients. Diane forms alliances with the health department, increases media awareness of the infection, and persuades the Canadian government to address its hidden public health problem.

1989
At a National Institutes of Health scientific meeting, James Miller presents a slide of a *Bb* bacterium half in and half out of a cell, demonstrating it may be intracellular. This would mean some classes of antibiotics frequently used in treating LD are unable to eradicate the infection from patients and may explain why diagnostic blood tests are not always accurate.

1989
Vera Preac-Mursic advances two theories as to why the immune system's antibody response to *Borrelia burgdorferi* can result in a negative test. One hypothesis is that

the bacterium may hide from the body's defense system; alternatively, the bacterium may bind to human antibodies, leaving too few free-floating antibodies left to measure.

Source: Preac-Mursic et al. (1989)

1990
The Council of State and Territorial Epidemiologists, funded by the Centers for Disease Control and Prevention, establishes the first uniform case definition for Lyme disease for reporting purposes. Unfortunately, the definition is narrow and is based solely on signs that a physician can see and measure. Symptoms communicated by the patient to the physician are discounted.

Source: Craven and Miller (1990)

1990
Researchers discover *Bb* in the lone star tick in the Midwest.

Source: Feir and Reppell (1990)

1990
The Australian government funds a study to look for *Bb* in ticks in Australia. Although ninety-two of twelve thousand ticks prove to have spirochete-like organisms, the government declares that *Bb* does not exist in Australia.

1990
The provincial government of British Columbia, Canada, claims that LD has not been established in British Columbia, despite the many personal reports to the contrary.

Source: Farley et al. (1990)

1991
Bb is demonstrated to have penetrated the human endothelial cells in the umbilical cord, proving both its intracellular capability and the potential threat to the developing fetus.

Source: Ma et al. (1991)

1991
The Lyme Borreliosis Foundation changes its name to the Lyme Disease Foundation, because the original name was so difficult to pronounce. The foundation's mission is broadened to include all tick-borne disorders.

1991
Seeds of activism in Australia start. Patient Terry Moore is finally diagnosed with Chronic Fatigue Syndrome after a long period of suffering. The LDF's international conference material offers her many insights into the real cause of her symptoms.

1991
Bb's ability to invade and survive in macrophages, an important component of the human immune system, is described.

Source: Montgomery et al. (1991)

1992
Led by activist Diane Kindree, the Vancouver, British Columbia, LBS persuades the B.C. Laboratory Centres for Disease Control to undertake another study. Within 4 months, a scientific team led by Dr. S. Banerjee finds the LD pathogen in ticks and rodents in twenty-five of twenty-seven sites investigated.

1992
Bela Bózsik establishes the Lyme Borreliosis Foundation of Hungary.

1992
Bb is shown to be able to invade skin fibroblasts, important cells found in connective tissue, and to be protected from ceftriaxone, a widely used intravenous antibiotic.

Source: Klempner et al. (1992)

1992
A proposal is advanced to divide the expanding number of strains of *Bb* into subgroups. The umbrella category is called *Borrelia burgdorferi* sensu lato. The subgroups are divided into *Borrelia burgdorferi* sensu stricto, *Borrelia garinii, Borrelia afzelii,* and eventually *Borrelia japonica* and *Borrelia miyamoto.*

Source: Baranton et al. (1992)

1993
Bb is successfully cultured from ticks in British Columbia, and the government declares the province "endemic" for the Lyme disease-causing spirochete.

Source: Banerjee (1993b)

1993
Terry Moore of Dee Why, Australia, establishes the Tick Alert Group Support (TAGS), the first nonprofit organization in Australia to raise awareness of LD, educate physicians and the general public, encourage scientific research, and provide support for patients. Bernie Hudson, a microbiologist at Royal North Shore Hospital in Sydney, Australia, provides scientific support and guidance.

A 3-year government research study fails to find *Bb* in Australian ticks.

1993
James Oliver and his colleagues from Georgia Southern University prove that the newly named *Ixodes dammini* tick (deer tick) is actually the already-known *Ixodes*

scapularis (black-legged tick). Their findings prove that the tick that carries *Bb* can be found across the country. This improves the ability of patients in the South and Midwest to be properly diagnosed in states where the existence of LD was once denied.

Source: Oliver, Jr. et al. (1993)

1993
S. Banerjee cultures spirochetes from an *Ixodes* tick species—*Ixodes angustus*—expanding the range of ticks that transmit the disease.

Source: Banerjee et al. (1994)

1994
Lyme borreliosis becomes a reportable disease in the province of British Columbia.

Source: Banerjee et al. (1994)

1994
The LDF establishes the *Journal of Spirochetal and Tick-borne Diseases,* the first international peer-reviewed journal dedicated to LD, other spirochetal and tick-borne disorders.

1994
W. F. Marshall and his colleagues from the Harvard School of Public Health conduct DNA amplification tests on white-footed mice preserved by a Massachusetts researcher in 1894. *Bb* is found, establishing the presence of the Lyme bacterium in the United States for at least 100 years.

Source: Marshall et al. (1994)

1995
David Dorward and Claude Garon, researchers at the Rocky Mountain Laboratories, create a videotape taken through an electron microscope that shows *Bb* invading a cell of the human immune system, dividing, and then breaking down the cell wall, causing the membrane to collapse around itself. The membrane serves as a cloaking device that may prevent the immune system from recognizing the presence of a foreign invader and mounting an antibody response.

This finding is presented at the LDF's international conference, held in British Columbia with the support of the B.C. Centres for Disease Control and the Vancouver, B.C., Lyme Borreliosis Society, Rocky Mountain Laboratories, and the Max von Pettenkofer Institute in Germany.

1995
The *Bb* infection is found in ticks collected and preserved from animals in Hungary and Austria more than a century earlier. Scientists from the University of Berlin in

Germany and the Harvard School of Public Health in the United States deserve the credit.

Source: Matuschka et al. (1995)

1996
Researcher Michelle Wills discovers *Bb* DNA by polymerase chain reaction in Australian ticks.

1996
Vera Preac-Mursic tests twenty strains of the Lyme bacterium and proves that extremely similar strains have widely different responses to various antibiotics. Sometimes, oral antibiotics work better than intravenous antibiotics, and sometimes the reverse. Similarly, some antibiotics work exceedingly well with one strain, but are ineffective with another strain that is almost identical. Preac-Mursic recommends that physicians consider using dual antibiotic therapy and treat patients for longer than just a few weeks.

Source: Preac-Mursic et al. (1996)

1996
Researchers announce a new *Borrelia* has been identified in lone star ticks in Maryland.

Source: Armstrong et al. (1996)

1996
A team of researchers led by a physician from the University of Texas discovers a new *Borrelia* in lone star ticks from New Jersey, New York, Missouri, and Texas. These prove difficult to culture, but use of polymerase chain reaction DNA testing proves they are new.

Source: Barbour et al. (1996)

1996
The antibiotic ceftin becomes the first drug approved for use in early Lyme disease.

1997
SmithKline Laboratories announces its vaccine trial results—that the vaccine is protective against the Lyme bacterium.

Appendix II
OFFICIAL CDC DISEASE DEFINITIONS

LYME DISEASE

In order to have a uniform description of Lyme disease for research purposes and to monitor the incidence of disease in different parts of the country, the Centers for Disease Control and Prevention (CDC) developed a very specific and narrow definition of Lyme disease. While valuable, this definition undoubtedly overlooks some cases and may differ somewhat from the cases your physician uses for a clinical diagnosis. This is for reporting purposes only.

Clinical Description

A systemic, tick-borne disease with ever-changing manifestations, including dermatologic, rheumatologic, neurologic, and cardiac abnormalities. The best clinical marker for the disease is the initial skin lesion, erythema migrans (EM), although this does not occur in all patients. A known tick bite is not required.

Case Definition

A case is confirmed if one of two criteria are met:

1. The characteristic erythema migrans rash is present.

or:

2. The patient has at least one late manifestation, as defined below, and laboratory confirmation of infection.

Erythema Migrans

For surveillance purposes, EM is defined as a skin lesion that typically begins as a red macule or papule and expands over a period of days to weeks to form a large round lesion, often with partial central clearing. A solitary lesion must reach at least 5 centimeters in size. Secondary lesions may also occur. Lesions occurring within several hours of a tick bite are signs of hypersensitivity and do not qualify as EM.

For most patients, the expanding EM lesion is accompanied by other acute symptoms, particularly fatigue, fever, headache, mild stiff neck, arthralgia, or myalgia. These symptoms are typically intermittent. The diagnosis of EM must be made by a physician. Laboratory confirmation is recommended for persons who are not known to have been exposed to ticks.

Late Manifestations

Criteria for late manifestation of Lyme disease include any of the following when an alternate explanation is not found:

Musculoskeletal system: Recurrent, brief attacks of objective joint swelling in one or a few joints, sometimes followed by chronic arthritis in one or a few joints. Manifestations that are not considered diagnostic criteria include chronic progressive arthritis that has not been preceded by brief attacks and chronic symmetrical polyarthritis. Arthralgia, myalgia, or fibromyalgia syndromes alone are not criteria for musculoskeletal involvement.

Nervous system: Any of the following, alone or in combination, are considered late neurological manifestations of Lyme disease: lymphocytic meningitis; cranial neuritis, particularly bilateral facial palsy; radiculoneuropathy; or, rarely, encephalomyelitis. Encephalomyelitis must be confirmed by showing *Borrelia burgdorferi (Bb)* antibody production in the cerebrospinal fluid (CSF) that is higher than serum levels. Headache, fatigue, tingling sensations, or mild stiff neck alone are not criteria for neurologic involvement.

Cardiovascular system: Acute onset, high-grade atrioventricular conduction defects that resolve in days to weeks and are sometimes associated with myocarditis. Palpitations, bradycardia, bundle branch block, or myocarditis alone are not criteria for cardiovascular involvement.

Laboratory Criteria for Diagnosis

Any one of these three criteria constitutes laboratory confirmation:

1. *Bb* is isolated from tissue or body fluid.
2. Diagnostic levels of IgM and IgG antibodies to *Bb* in blood serum or cerebrospinal fluid are demonstrated.
3. Significant change occurs in IgM or IgG antibody response to *Bb* in serum samples studied during the acute and convalescent phases of disease.

Each of the 50 states is allowed to determine the criteria for laboratory confirmation and diagnostic levels of antibody. Syphilis and other known causes of false-positive serologic test results should be excluded if laboratory confirmation is based on serologic testing alone.

ROCKY MOUNTAIN SPOTTED FEVER

Clinical Description

An illness most commonly characterized by acute onset and fever, usually accompanied by myalgia, headache, and, in about two-thirds of the cases, a characteristic rash on the palms and soles.

Laboratory Criteria for Diagnosis

Any one of these three criteria constitutes laboratory confirmation:

1. Fourfold or greater rise in antibody titer to the spotted fever group antigen by any of the following laboratory tests—immunofluorescent antibody (IFA), complement fixation (CF), latex agglutination (LA), microagglutination (MA), or indirect hemagglutination (IHA) test. Alternatively, criteria for a single titer test have been established.
2. Demonstration of positive immunofluorescence of skin lesion in a biopsy or organ tissue in an autopsy.
3. Isolation of *Rickettsia rickettsii* from a clinical specimen.

Case Classification

The CDC classifies a case as "probable" Rocky Mountain spotted fever on the basis of supportive serological tests.

TULAREMIA

Although the CDC stopped collecting surveillance information about tularemia in 1995, it is useful to have the official case definition as a guideline for understanding this tick-borne disease.

Clinical Description

Tularemia may take several distinct forms:

1. Ulceroglandular—cutaneous ulcer with regional lymph node involvement
2. Glandular—regional lymph node involvement without ulcer

3. Oculoglandular—conjunctivitis with lymph node involvement in front of the ear
4. Intestinal—intestinal pain, vomiting, and diarrhea; the throat may be inflamed
5. Pneumonic—lung involvement
6. Typhoidal—feverishness without early local signs and symptoms

The clinical diagnosis is supported by evidence or history of a tick or deerfly bite, exposure to tissues of a mammalian host of *Francisella tularensis,* or exposure to potentially contaminated water.

Laboratory Criteria for Diagnosis

Any one of these three criteria constitutes laboratory confirmation:

1. Isolation of *F. tularensis* from a clinical specimen, or
2. Demonstration of *F. tularensis* in a clinical specimen by immunofluorescence, or
3. Fourfold or greater rise in antibody titer between the acute and convalescent phases of disease in serum specimens obtained at least 2 weeks apart and analyzed at the same time and in the same laboratory

Case Classification

A case is classified as "probable" tularemia if clinical symptoms are present and antibody levels in a serum sample obtained after the onset of symptoms support the diagnosis. A case that is verified by a laboratory is considered "confirmed."

Appendix III

RESOURCES OF THE LYME DISEASE FOUNDATION

Lyme Disease Foundation, Inc.
One Financial Plaza
Hartford, Connecticut 06103
(860) 525-2000

Electronic Mail:
Lymefnd@aol.com
104477.3111@compuserve.com

Web site: http://www.lyme.org
24-hour Hotline: (800) 886-LYME

The Lyme Disease Foundation is the first, largest, and most influential national health care agency dedicated to finding solutions to tick-spread disorders. The LDF does this through education, research, and advocacy programs. The board of directors includes a congressman, a health commissioner, business leaders, physicians, and the individual who discovered the Lyme disease bacterium. The list of scientific advisors to the Lyme Disease Foundation reads like a Who's Who in tick-borne and spirochetal diseases, and strong ties exist to the international scientific community.

A $30 donation entitles you to a 1-year subscription to the bimonthly newsletter *LymeLight*. The Lyme Disease Foundation is a member agency of the United Way, Combined Health Appeal, Combined Federal Campaign, and various state campaigns. To donate write "LDF 860-525-2000" on your workplace designation card.

These are among the key activities of the Lyme Disease Foundation:

- Sponsors an annual, medically accredited, international scientific conference
- Publishes the *Journal of Spirochetal and Tick-borne Diseases,* a peer-reviewed scientific publication that covers new scientific research about Lyme disease and other tick-borne and spirochetal diseases
- Encourages professional networking for researchers and physicians
- Provides physician referrals for patients
- Educates state and federal elected officials and advocates greater attention to Lyme disease

- Operates a 24-hour computerized telephone hotline used by seventy-thousand people every year
- Sponsors chat rooms in cyberspace for patients and health care professionals (see appendix IV)
- Sponsors task forces to conduct education and other services at the state level

EDUCATIONAL MATERIAL

In an effort to educate both the public and the medical community, the Lyme Disease Foundation produces a variety of publications, including brochures, newsletters, and a scholarly journal. In addition, educational brochures, posters, bumper stickers, and other materials are available. Contact the foundation for prices on individual items and for bulk purchases.

Some of the videotapes and other materials may also be available at your local library. If not, ask your local library to contact the state library to find out what Lyme disease material is available on interlibrary loan.

Web Site

The foundation's Web site on the Internet has a full range of disease information and addresses of state health departments and insurance departments. It lists state libraries with free Lyme disease educational information and provides links to government pages and international resources.

Newsletter and Scientific Journal

LymeLight newsletter has information written for a general audience. It is published six times a year.

The *Journal of Spirochetal and Tick-borne Diseases* is a peer-reviewed quarterly devoted to all aspects of a variety of diseases. It includes color pictures.

Brochures and Cards

Frequently Asked Questions about Lyme Disease: Brochure provides an overview of testing, treatment, and pregnancy-related issues.

A Guide to Lyme Disease: Brochure describes Lyme disease, the Lyme Disease Foundation and the material available through the foundation. Pictures are in color.

Lyme Disease and Pets: Brochure outlines Lyme disease symptoms and treatment in various animals.

A Guide to Tick-borne Disorders: Brochure describes symptoms, diagnosis, and treatment for a variety of diseases.

Tick Alert Card: Easy-to-carry card describes ticks, tick removal techniques, and symptoms of Lyme disease.

Informational Packets

Lyme Disease Awareness Packet: Educational, letter-size posters, brochures listed above, case counts, information in Spanish, information about insurance issues. Includes a poster of general diagnostic information.

Lyme Disease Scientific Packet: Copies of a variety of scientific articles, including testing and treatment issues. Includes a poster of scientific diagnostic information.

Scientific Conference Compendium: Articles by speakers from the Lyme Disease Foundation's recent scientific conference.

Videotapes

Lyme Disease: Facts for Kids (18 minutes apiece)**:** Two educational puppet videos, available in English, Spanish, and open-captioned for the hearing impaired, featuring Emilio Delgado ("Luis" from Sesame Street). These videos were the silver plaque winners of the 1994 International Communication Film and Video Festival. *Dr. Ticked-Off and His Tick Patrol* (grades K-3) features a visit to the office of Dr. Ticked-Off. *WTIK News Flash: DO A TICK-CHECK!* (grades 4–8) is set in the fictional studio of WTIK.

Lyme Disease: What You Should Know (60 minutes): A comprehensive video designed to inform adults about Lyme disease. Interviews with more than thirty physicians, researchers, patients, health department officials, and school officials provide in-depth information.

How to Establish and Conduct a Lyme Disease Self-Help Group: A step-by-step guide to launching a healthy Lyme disease self-help group. Packaged with two how-to booklets, a start-up video, member brochure, and other supporting material.

Lyme Disease: Diagnosis and Treatment (60 minutes): Four physicians discuss the challenges of diagnosing and treating Lyme disease.

Slide Shows and Posters

Community Education Slide Show: 20 slides with script.

Lyme Disease Scientific Slide Show: 40 color slides with script.

Lyme Disease Diagnostic Picture Poster Set: This poster set includes 14 color pictures and is great for a physician's office, school, hospital, or community group.

Posters describe symptoms, both by stage and organ system, and provide other facts about Lyme disease.

Lyme Disease Prevention Poster: Five jumbo pictures describe Lyme disease, prevention techniques, and major symptoms.

Integrated Educational Packages

Lyme Disease Community Education Program: An educational package containing the *Facts for Kids* and *What You Should Know* videotapes, the *Community Education* slide show and script, two full-color diagnostic posters, original brochures and handouts, bumper stickers, and a lightweight tabletop color display. A yellow zip-top tote bag for carrying or shipping is included.

Lyme Disease Scientific Investigator Program: This hands-on program uses Lyme disease as an example to teach scientific methods to junior high school students. An instructional videotape called *Lyme Disease: An Investigative Survey on Personal Prevention* is included, along with a 1-day and a 4-day lesson plan for teachers, survey forms, and Lyme disease brochures.

Workplace Education Program: Educates employees, their families, and other groups about Lyme disease and other tick-borne diseases. Included are: two 20-minute educational videos (one is about the program, the other is a self-contained educational video about ticks, disease symptoms, and prevention techniques), employer implementation handbook, payroll inserts, posters, and the *Community Education Slide Show.*

The Works: Contains the complete *Community Education Program, Workplace Education Program, Lyme Disease: Diagnosis and Treatment* videotape, *Lyme Disease: Facts for Kids* (bilingual), and the *Lyme Disease Scientific Packet.*

Bumper Stickers and Buttons

Lyme Disease: A Ticking Bomb: Time bomb graphic (bumper sticker)

A Tick in Time Prevents Lyme!: Tick removal graphic (bumper sticker)

Your Brain on LD: Cheese vs Swiss cheese graphic (bumper sticker)

I Support Lyme Disease Education and Research: (button)

Ask Me About Lyme Disease!: (specify button or bumper sticker)

Clothing

Sports Mesh Cap—Dr. Ticked-Off *DO A TICK CHECK!*

Sweatshirt—*LDF—Unmasking the Great Masquerader:* Cuddly, fleece-lined long-sleeved sweatshirt. Available in forest green or blue. Sizes medium, large, and extra large.

Appendix IV

TICK BYTES IN CYBERSPACE

Not yet convinced of the advantages of going online to learn more about Lyme disease? Here are some of the things you can do:

- You had a tick bite and want to identify the tick. Log on to the World Wide Web and find pictures of a wide variety of ticks.
- Your mother has just been diagnosed with Lyme disease and wants to talk with someone else who has the disease, but she is homebound. Help her join an online support group.
- You are a physician who has just diagnosed the first case of ehrlichiosis in your area. You want to read up on the disease and talk to researchers who are studying it. Go online, conduct a medical search, then ask a world expert a question.
- You are an Australian researcher looking for technical information. Join a research chat group, hosted by a Canadian public health official, and find out where to get a culture medium.
- Your daughter is writing a school paper on babesiosis and needs more information. Send her online for material she can download, including government graphs on the incidence of disease.
- You can find out the name and electronic mail address of your congressional representatives so you can urge them to allocate more federal funds for research into tick-borne diseases.
- You hear that there is a startling new development in Lyme disease. Go online, read the article in cyberspace, and print out a copy to show a friend.

Here are the places to find the information you want:

HOME PAGES RELATED TO LYME DISEASE

Lyme Disease Foundation

http://www.lyme.org

This Web site contains everything you need to get started on your search for information. It explains Lyme disease and other tick-borne disorders in detail, provides pictures of ticks and the clinical manifestations of disease, and links to other sites around the world. You can review thousands of Lyme-related medical articles and their abstracts.

Lyme Disease Network of New Jersey

http://www.lymenet.org

A great site for basic background. The site explains the disease and its transmission, contains sample treatment protocols, and provides background on relevant legal decisions. Links are provided to newsgroups and to locations with pictures of ticks and rashes. There is a search capability to review thousands of Lyme-related medical articles.

Lyme Resource

http:// www.sky.net/~dporter/lyme1.html

Basic information about the disease, with links to other sites.

Colorado State University

http://www.ColoState.edu/Depts/Entomology

(This address is case sensitive, so type it exactly as is)

Select "entomology." An extraordinary set of links to other Web pages, newsgroups, gopher search sites, bulletin boards, periodicals, databases, and even a pesticide newsletter.

The Pasteur Institute in France

http://www.pasteur.fr

An outstanding source of information on the microbiology of spirochetes. Included is the registration of *Bb* strains, substrains, and genetic codes. Information on other spirochetes and tick-spread diseases is available.

The University of Rhode Island Tick Research Lab

http://www.uri.edu/artsci/zool/ticklab

General information on Lyme disease, babesiosis, and ehrlichiosis. Includes hand-drawn pictures of various ticks and links to other Web sites.

Iowa State University

http://www.ent.iastate.edu/ImageGallery

Great site to view pictures and brief videotapes of ticks and tick dissection, mosquitoes, lice, beetles, and other creatures.

Lymenet Newsletter

http://www.lymenet.org

Informative electronic mail newsletter that is well worth a subscription. You can also contact them through the mail at 1050 Lawrence Avenue, Westfield, New Jersey 07090-3721, or by telephone at (908) 789-7346.

European Union Concerted Action on Lyme Borreliosis (EUCALB)

http://www.dis.strath.ac.uk/lymeeu/index.htm

This is a site for the European scientific community to document information on European LB and their progress toward standardization of medical terms, concepts, and research.

CHAT GROUPS

Chat groups are live conversations reminiscent of telephone party lines. You can talk either one-to-one or in group meetings. Most chat groups are found on commercial online services, but some chat sites are beginning to appear on the Internet.

America Online

America on Lyme: Monday from 8-9 P.M., EST.
Key word: "Better Health" or "Pen"
Location: Equal Access Cafe
Hosts: PENKvetch and PENSuz

Patient and family support group. Typically, a guest speaker is scheduled, often followed by a chat in a private room.

Compuserve

Patient/public chat: Go "Pub/hlth." Meets the second Wednesday of the month from 9:30 P.M. to 10:30 P.M. EST. Hosted by Roberta Bethencourt, M.S.W., a social worker in private practice. This is a freewheeling discussion group.

Health care professional chat: Go "Pub/hlth." Meetings are the third Wednesday of the month from 9:30 P.M. to 10:30 P.M., EST. Hosted by Julie Rawlings, M.P.H., an epidemiologist with the Texas Department of Health, and Phillip Watsky, M.D., a Connecticut physician in private practice. Attendees must be scientists, doctors, researchers, nurses, and so on. To obtain the "key" to attend, e-mail the Lyme Disease Foundation (104477.3111@compuserve.com) at least one day before the meeting. Health care professionals are also encouraged to attend the public chat the week before.

NEWSGROUPS

sci.med.diseases.lyme

An international newsgroup based in Scotland that fosters information exchanges among patients, caregivers, and medical professionals about prevention, symptoms, and treatment. This is a supportive newsgroup that encourages open communication, especially among Lyme disease sufferers.

sci.bio.entomology.misc

A relatively new international newsgroup where entomologists can post questions and answers.

FEDERAL GOVERNMENT HOME PAGES

Department of Health and Human Services (DHHS)

http://www.dhhs.gov

DHHS home pages, with links to the National Institutes of Health and the Centers for Disease Control and Prevention.

National Institutes of Health (NIH)

http://www.nih.gov

The NIH Web pages include information about grants, research projects, ongoing activities, press releases, and more.

Centers for Disease Control and Prevention (CDC)

http://www.cdc.gov

Provides information about the mission of the CDC, grants, disease statistics, reportable disease case definitions, and other information.

The White House

http:// www.whitehouse.gov

The President's e-mail address is President@whitehouse.gov

Information about events in the White House. It includes daily press briefings and information from the president and vice president.

U.S. Senate

http://www.senate.gov

Provides information about the Senate, including its members, committee structure, leadership, and support offices. Links are provided to other government sources of information, including the House of Representatives, the Library of Congress, the Government Printing Office (which provides copies of legislation), the General Accounting Office, and the White House. You can also consult Thomas, a computer program that provides legislative information on the Internet.

U.S. House of Representatives

http://www.house.gov

A superior list of issues in the House, including schedules of events, pending legislation, and committee memberships. A complete list of names and addresses of House members is available.

SEARCH SITES

Search engines allow you to search for other relevant information, including any Web sites, chat groups, or newsgroups that spring up after publication of this book. The following search sites are among the best for people seeking information on Lyme disease and other tick-borne disorders.

Santel

http://www.santel.lu/Santel/diseaseslyme.html

Multiple medical searches around the world.

Webcrawler

http://www.webcrawler.com

Especially useful for searching Internet Web pages.

YAHOO

http://www.yahoo.com

Easy-to-use search engine for items that are indexed by subject or can be searched by a key word. You can also search for Internet chat sites.

Nexor

http://nexor.co.uk

Select "public" to be transferred to the World Wide Web and a wide variety of Archie searches for information. Included is a list of specialized international Archie sites.

INFOSEEK

http://www2.infoseek.com

A Web and newsgroup search engine that uses either key words or categories.

EXCITE Search

http://www.excite.com

This is useful for searching Web sites, newsgroups, and press releases. You can choose a category or key word to get a list of recommended sites with a description of their contents.

Net Vet at Washington University

http://netvet.wustl.edu

Extraordinary list of veterinary information. Lists worldwide sites, meeting conferences, veterinary publications, laws, and regulations.

Appendix V
OTHER RESOURCES

ASSOCIATIONS AND SUPPORT GROUPS—UNITED STATES[1]

Arizona
1821 N. 87th Way
Scottsdale, AZ 85257
(602) 994-5449

Meets monthly at Scottsdale Public Library. Times vary, so call for further information.

California
Trinity County Lyme Disease Network
Box 707
Watsonville, CA 96093

Meets at Nazarene Church Education Annex on the third Thursday of every month from 10 to 11:30 A.M.

Connecticut
Bristol-Burlington Support Group
4 Short St.
Bristol, CT 06010

Meets at the First Congregational Church, Maple Street, Bristol, on the third Wednesday of the month at 7 P.M.

Connecticut Self-Help
667 Route 169
Woodstock, CT 06281
(860) 928-9702

Meets at the Putnam Public Library in Windham on the first Saturday of each month from 1 to 3 P.M.

Illinois
Quad Cities Lyme Disease Information Support Group
505 East 22nd Avenue
Coal Valley, IL 61240

Meets at Trinity Medical Center in Moline on the third Thursday of every month from 7 to 9 P.M.

---

[1]Unless a telephone number is provided, these groups should be contacted by mail only.

Iowa
Iowa Lyme Disease Association
Box 291
Brighton, Iowa 52540
(319) 353-4529

Support group meets at the Coralville Library the second Saturday of the month from 10 A.M. to noon.

Maryland
Washington County Lyme Disease Support Group
(301) 739-3250

Support group meets monthly. Call the Washington County Free Library for a schedule.

Central Maryland Lyme Disease Support Group
506 Granleigh Road
Owings Mills, MD 21117

Support group meets at the Northwest Baptist Church, 300 Westminster Rd., Reistertown, on the second Thursday of the month from 7:30 to 9:30 P.M.

Hartford County Lyme Disease Support Group
Box 13
Street, MD 21154
(410) 452-5043

Provides support for local residents. Meets at the Highland Presbyterian Church in Street on the second Thursday of the month from 7:30 to 9 P.M.

Massachusetts
Western Massachusetts Lyme Disease Awareness Association
49 Cherry Street
Easthampton, MA 01027
(413) 529-0343

Support group meets on the first floor of the Holyoke Hospital meeting room in Hamden on the third Thursday of the month from 7 to 9 P.M.

Lyme Disease Self-Help Group of Greater Boston
Contact them through the Lyme Disease Foundation

Meets at the Islington Branch of the Westwood Library at 280 Washington Street in Westwood on the first Thursday of every month from 7 to 9 P.M.

Michigan
Lyme Alliance of South Central Michigan
Box 454
Concord, Michigan 49237

Support group meetings are held at the Concord United Methodist Church, 113 S. Main Street, Concord, on the third Tuesday of the month from 7 to 8:30 P.M.

Minnesota
Duluth/Superior Lyme Disease Support Group
902 Grandview Avenue
Duluth, Minnesota 55812
(218) 728-3914
71604,1030@compuserve.com

Support group provides information about Lyme disease and support. Meetings are held at St. Luke's East Clinic Building, Room 227, 1001 East First Street, Duluth, on the first Thursday of the month at 7 P.M.

New Jersey
Morris Area Lyme Support Group
Box 1483
Morristown, New Jersey 07960-3610

Self-help group meets the third Tuesday of the month at the Presbyterian Church's Parish House, 65 South Street, Morristown, 7 to 9:30 P.M.

New York
Long Island Lyme Association
Box 1847
North Massapequa, New York 11758
(516) 797-LYME

Support group meets at the Mid-Island Hospital at 4295 Hempstead Turnpike, Bethpage, on the third Monday of every month at 7:30 P.M. The association also provides support and information to families.

Lyme Borrelia Outreach
Box 496
Mattituck, New York 11952
(516) 298-9606

Support group meets at the Riverhead Free Library, 330 Court Street, Riverhead, on the second Tuesday of the month at 6:45 P.M. The organization also produces a public access cable television program on Lyme disease and issues a bimonthly newsletter. Ask your local access cable station to run their program.

Lyme Disease Coalition of New York and Connecticut
Box 463
Katonah, New York 10536

Support group meets at the First Presbyterian Church, 31 Bedford Rd., Katonah, at 11 A.M. on the third Saturday of the month.

North Carolina
Lyme Help for North Carolina
Box 61284
Raleigh, North Carolina 27661
(919) 217-9896

The self-help support group meets once a month on Saturdays. Meeting locations vary, so call for further information.

Ohio

Greater Cleveland Lyme Disease Support Group
7644 Main Street
Cleveland, Ohio 44138
(216) 235-4163

Support group meets at Southwest General Health Center on the third Sunday of each month from 3 to 5 P.M. A subscription to a newsletter called *Bull's Eye* is available for a $5 donation.

Lyme Disease Association of Ohio
Box 751511
Dayton, Ohio 45475
(513) 436-0267
rparrett@IBM.net
Lymewife@aol.com

The association is involved in education and advocacy work to support Lyme disease patients, their health care providers, and others at risk in Ohio. Publishes a newsletter called the *Lyme Disease Reporter,* available for a $15 donation.

Oregon

Northwest Lyme Disease Support Network
Good Samaritan Hospital
1015 N.W. 22nd Avenue, N-300
Portland, OR 97210
(503) 292-5271

Meets at the hospital bimonthly on Sundays from 2 to 4 P.M. Call for schedule.

Pennsylvania

Lyme Disease Community Coalition (LDCC)
401 Baker Lane
Coatsville, Pennsylvania 19320
(610) 384-9622

The coalition emphasizes improving the quality of life and understanding for Lyme disease sufferers and their families. Support group meets at the Brandywine Health Pavillion, Oaklands Corporate Center, Exton, on the third Tuesday of the month from 6:45 to 8:45 P.M.

Westmoreland County Lyme Disease Association
164 Bonita Drive
Greensburg, PA 15601
(412) 836-8104

The association was established to provide financial assistance and support for local patients.

Grove City Lyme Disease Information and Support Group
98 Whitaker School Road
Grove City, PA 16127
(412) 458-0509

Meets at Pine Grove Community Center on the last Sunday of each month from 2 to 4 P.M.

Rhode Island

Rhode Island Lyme Network
46 Barnes Street
Providence, RI 02906
(401) 274-5921

Contact this new self-help group for further information.

Wisconsin

Burnett County Lyme Disease Support Group
28749 Barley Road North
Danbury, WI 54830
(715) 656-7138

Meets at the Burnett County Library in Webster on the second Thursday of every month from 7 to 9 P.M.

ASSOCIATIONS AND SUPPORT GROUPS—INTERNATIONAL

Australia

Tick Alert Group Support, Inc. (TAGS)
Box 1551
Dee Why, N.S.W. 2099
Tel: 02 9971 4621
Medical Advisor: Dr. Bernie Hudson (Fax: 437 5746;
e-mail bhudson@blackburn.med.su.oz.au)

An excellent advocacy and support group. Dr. Bernie Hudson, one of Australia's leading Lyme researchers, keeps TAGS in close contact with the medical community.

Canada
British Columbia
Vancouver, B.C., Lyme Borreliosis Society
Box 91535
West Vancouver, British Columbia V7V 3P2
(604) 922-3704

This organization has made significant progress in pushing Canada to recognize Lyme disease and has established an excellent relationship with the B.C. Centres for Disease Control. The *LYME NEWS* newsletter is an excellent resource, available for a donation of $25 (in Canada) or $35 (outside Canada).

Manitoba
Lyme Borreliosis Support Group of Manitoba
350 Rougeau Avenue
Winnipeg, Manitoba R2C 4A2
(204) 663-3570
(204) 347-8042

This self-help group helps increase awareness and understanding of Lyme Borreliosis and other tick-spread diseases. Contact them for local meetings.

Ontario
Lyme Disease Association of Ontario
365 Saint David Street South
Fergus, Ontario N1M 2L7
(519) 843-3646

Self-help meetings at various locations throughout the province. A $25 donation provides four issues of its newsletter *Lyme Alert.*

Hungary
Lyme Borreliosis Foundation
Box 64 - Gyáli St. 2–6
Budapest, Hungary
H-1966
Phone 011-361-215-2250 ext. 226
Fax 011-361-215-0148

Established in 1991 by Dr. Béla Bozsik, this organization provides medical education, patient support, and media education.

GOVERNMENT RESOURCES

State Health Departments
Most states have health departments that may be able to provide you more informa-

tion about the risks of Lyme disease and other tick-borne disorders in your area. Consult the government pages of your telephone book for this information. It may be listed as the Department of Health, the Department of Public Health, the Department of Health and Social Services, the Department of Health and the Environment, or something similar.

Centers for Disease Control and Prevention
For Lyme disease:
Division of Vector-borne Infectious Diseases
Box 2087
Fort Collins, Colorado 80522
(303) 221-6453

For babesiosis and ehrlichiosis:
U.S. Department of Health and Human Services
Atlanta, Georgia 30333
(404) 639-3311

National Institutes of Health
Lyme Lines
National Institute of Allergy and Infectious Diseases
Box AMS
9000 Rockville Pike
Bethesda, Maryland 20891

Contact NIH for publications and information about ongoing Lyme disease research projects.

LABORATORY TESTING

BBI—North American Clinical Labs
75 North Mountain Road
New Britain, Connecticut 06053
(800) 886-6254
(860) 225-1900

This laboratory is well-equipped to detect disease-causing pathogens in both ticks and human beings. Reports all bands on Western blot tests.

IGeneX, Inc. Reference Laboratory
797 San Antonio Road
Palo Alto, California 94303
(800) 832-3200

Another well-equipped laboratory suitable to detect disease-causing pathogens in both ticks and human beings. Reports all bands on Western blot tests.

New Jersey Laboratories, Inc.
1110 Somerset Street
New Brunswick, New Jersey 08901
(908) 249-0148

This laboratory will test ticks for the pathogens that cause Lyme disease and human granulocytic ehrlichiosis.

Connecticut Veterinary Diagnostic Testing Labs
University of Connecticut
Department of Pathobiology
61 North Eagleville Road U-203
Storrs, Connecticut 06269
(860) 486-0808

This laboratory performs a variety of tests for Lyme disease for domestic pets, livestock, and ticks.

Veterinary Research Associates
10 Executive Boulevard
Farmingdale, New York 11735
(516) 753-4100

Provides a variety of laboratory tests for tick-borne disorders for domestic pets and livestock.

TICKS AND TICK PICTURES

Ward's Natural Science Establishment
Box 92912
Rochester, New York 14692
(800) 962-2660

Various tick samples mounted on slides are available at a cost of about $10 apiece.

Microscopy Today
Box 122
Middletown, WI 53562
(608) 836-1970
e-mail:microtoday@aol.com

Enlarged electron microscope photographic close-ups of the *Ixodes scapularis* mouthparts are available. Ask for catalog #DS-03. $39.00

REPELLENTS

Avon
9 West 57th Street
New York, New York 10019
(800) FOR-AVON

Produces Avon's Skin-So-Soft Mosquito, Flea and Deer Tick Repellent Lotion, an FDA-approved Lyme tick repellent that is nontoxic and effective.

Amway
(800) 253-6500

Markets Hour Guard cream, a 33% deet concentration in a slowly evaporating polymer. Originally developed by 3-M as Ultrathon, Hour Guard repels ticks (with decreasing effectiveness) for about 12 hours and is designed to reduce skin absorption. A 2-ounce tube retails for about $12 and lasts for about twelve uses.

Coulston Products Inc.
Box 30
Easton, Pennsylvania 18044
(800) 445-9927

Sells tick-killing spray for use on clothing. Coulston also makes a product for use on dogs. These products are available in most lawn and garden stores.

Hartz Mountain
400 Plaza Drive
Secaucus, New Jersey 07094
(201) 481-4800

Produces pet products, including a full flea and tick repellent system that is effective against *Bb*-carrying ticks. These products are available at most grocery stores.

Mallinckrodt Veterinary
421 East Hawley Street
Mundelein, Illinois 60060
(847) 949-3300

Produces a fully integrated line of pet products for fleas and ticks. Call to find out if a veterinarian in your area carries this product line.

National Pesticide Telecommunications Network
6:30 A.M. - 4:30 P.M. (Pacific Time):
(800) 858-7378 (public inquiries)
(800) 858-7377 (medical professionals and government agencies)

Pesticide information on the use of pesticides and recognizing and managing poisons; toxicology/ symptomatic reviews, health and environmental effect reviews, and cleanup and disposal procedures.

TWEEZERS

Tick Tank
Ford V. Swick
551 Valley Road, Suite 129
Upper Montclair, New Jersey 07043

This company makes an excellent tick removal tweezers kit. The small circular kit hangs on a long rope so you can wear it around your neck whenever you are in tick-infected areas. This is a great gift for friends and family. Keep one in your car and another in your pocket.

Scandinavian Natural Health & Beauty Products, Inc.
13 North Seventh Street
Perkasie, Pennsylvania 18944
(215) 453-2505
(800) 688-2276

This company produces a spring-loaded clamp-type tweezers. Although some entomologists question its value, others swear by it. Personally, I don't recommend it, but you may decide you want to give it a try.

VACCINES

For information about ongoing vaccine research, you may want to contact these pharmaceutical companies:

Connaught Laboratories
Route 611, Box 187
Swiftwater, Pennsylvania 18370

SmithKline Beecham Pharmaceuticals
Clinical Research, Development & Medical Affairs
Four Falls Corporate Center
Route 23 and Woodmont Avenue
Box 1510
King of Prussia, Pennsylvania 19406

MedImmune, Inc.
35 West Watkins Mill Road
Gaithersburg, Maryland 20878
(301) 417-0770

Fort Dodge Animal Health
9401 Indian Creek
Overland Park, Kansas 66225
(800) 477-1365

INSURANCE AND HEALTH INFORMATION

State Insurance Departments
If you and your managed care organization cannot agree about what health care services should be covered, you may want to contact your state health insurance department for assistance or to file a complaint. Consult the government pages of your telephone book to find the insurance department in your state.

Medical Information Bureaus
Box 105
Essex Station
Boston, Massachusetts 02112
(617) 426-3660

and

330 University Avenue
Toronto, Ontario M5G 1R7
Canada
(416) 597-0590

Ask the Medical Information Bureau to provide a copy of any medical records it maintains on you. The MIB is used by insurance companies to determine your eligibility for health care and life insurance. If there are errors in your files, you will be given a chance to correct them.

Travel Medicine
351 Pleasant Street
Suite 312
Northampton, Massachusetts 01060
1-800-872-8633

If you are planning a trip anywhere in the world, call Travel Medicine first for information about various diseases and the appropriate precautions you should take.

Appendix VI

PERSONALIZED MEDICAL LOG

This log will allow you to keep track of any tick-related problems you or other members of your family experience. You should make copies of this blank form to fill out as necessary. Feel free to tape your tick right on the page as part of your permanent record.

DATE ______________________________

INFORMATION

Name of person bitten ______________________________

How long tick was attached______________________________

Place on body bitten______________________________

How tick was removed ______________________________

Type of tick ______________________________

Town where you were bitten ______________________________

Result of tick testing ______________________________

Appointments and phone calls to doctor______________________________

Your tests and results ______________________________

____

Symptoms ______________________________

____

____

Treatment ______________________________

____

____

SELECTED BIBLIOGRAPHY

Ackermann, R. B. Rehse-Küpper, E. Gollmer and R. Schmidt. (1987) "Chronic neurologic manifestations of erythema migrans borreliosis." *Annals of the New York Academy of Sciences,* Vol. 539, p. 16–23.

Ackermann, R., P. Hörstrup and R. Schmidt. (1984) "Tick-borne meningopolyneuritis (Garin-Bujadoux, Bannwarth)." *Yale Journal of Biology and Medicine,* Vol. 57, p. 485–490.

Altaie, S., S. Mookherjee, E. Assian, F. Al-Taie, S. M. Nakeeb, S. Siddiqui and L. Duffy. (1996) "Transmission of *Borrelia burgdorferi* from experimentally infected mating pairs to offspring in a murine model." *VII International Congress on Lyme Borreliosis.* San Francisco, CA: University of California, Berkeley, p. 287.

Angelov, L., T. A. Rakadjieva and T. R. Gancheva. (1996) "Non-transmission mechanism of infection in Lyme borreliosis." *VII International Congress on Lyme Borreliosis.* San Francisco, CA: University of California, Berkeley, p. 110.

Appel, M. (1990) "Lyme disease in dogs and cats." *The Compendium,* Vol. 12, No. 5, p. 617–626, 665.

Appel, M. and R. Jacobson. (1995) "CVT Update: Canine Lyme disease." *Kirk's Current Veterinary Therapy XII: Small Animal Practice.* Philadelphia, PA: W. B. Saunders, p. 303–308.

Appel, M., S. Allan, R. Jacobson, T. Lauderdale, Y. Chang, S. Shin, W. Thomford, R. Todhunter and B. Summers. (1993) "Experimental Lyme disease in dogs produces arthritis and persistent infection." *The Journal of Infectious Diseases,* Vol. 167, p. 651–664.

Armstrong, P., S. Rich, R. Smith, D. Hartl, A. Spielman and S. Telford, III. (1996) "A new *Borrelia* infecting lone star ticks." *Lancet,* Vol. 347, p. 66–67.

Background Information on deet. (1991) New York, NY: NYS Department of Health, Bureau of Toxic Substance Assessment, May 20, p. 1–22.

Bakken, J., S. Dumbler, S.-M. Chen, M. Eckman, L. Van Etta and D. Walker. (1994) "Human granulocytic ehrlichiosis in the upper midwest United States." *Journal of the American Medical Association,* Vol. 272, No. 3, p. 212–218.

Banerjee, S. (1993a) "Isolation of *Borrelia burgdorferi* in British Columbia." *Canada Communicable Disease Report,* Vol. 19–24, p. 204–205.

Banerjee, S. (1993b) "A search for the Lyme disease spirochetes in British Columbia." *British Columbia Provincial Laboratory News,* Vol. 2, No. 3 p. 3–4.

Banerjee, S., C. Stephen, K. Fernando, S. Coffey and M. Dong. (1996) "Evaluation of dogs as sero-indicators of the geographic distribution of Lyme borreliosis in British Columbia." *Canadian Veterinary Journal,* Vol. 37, p. 168–169.

Banerjee, S., M. Banerjee, J. Smith and K. Fernando. (1994) "Lyme disease in British Columbia—an update." *British Columbia Medical Journal,* Vol. 36, p. 540–541.

Banerjee, S., M. Banerjee, K. Fernando, M. Y. Dong, J. A. Smith and D. Cook. (1995) "Isolation of *Borrelia burgdorferi,* the Lyme disease spirochete from rabbit ticks, *Haemaphysalis leporispalustris* from Alberta." *Journal of Spirochetal and Tick-borne Diseases,* Vol. 2, No. 3, p. 23–24.

Baranton, G., D. Postic, I. S. Girons, P. Boerlin, J. Piffaretti, M. Assous and P. Grimont. (1992) "Delineation of *Borrelia burgdorferi* sensu stricto, *Borrelia garinii,* sp. nov., and group VS461 associated with Lyme borreliosis. *International Journal of Systematic Bacteriology,* Vol. 115, p. 91.

Barbour, A., G. Maupin, G. Teltow, C. Carter and J. Piesman. (1996) "Identification of an uncultivable *Borrelia* species in the hard tick *Amblyomma americanum*: Possible agent of a Lyme disease-like illness." *Journal of Infectious Diseases,* Vol.173, p. 403–409.

Barbour, A. G. (1984) "Isolation and cultivation of Lyme disease spirochetes." *Yale Journal of Biology and Medicine,* Vol. 57, p. 521–525.

Benach, J. L. (1983) "Spirochetes isolated from the blood of two patients with Lyme disease." *New England Journal of Medicine,* Vol. 308, p. 740.

Ben-ari, E. (1995) *Clinical Trial shows Minocycline is Safe and Effective for Rheumatoid Arthritis.* Bethesda, MD: NIAMS Research News Press Release, January 14, p. 1–3.

Berger, B., O. J. Clemmenson and A. Ackermann. (1983) "Lyme disease is a spirochetosis: a review of the disease and evidence of its cause." *American Journal of Dermopathology,* Vol. 5, p. 111.

Berger, B. (1989) "Cutaneous manifestations of Lyme borreliosis." *Rheumatic Disease Clinics of North America,* Vol. 15, No. 4, p. 627–634.

Berglund, J., R. Eitrem, K. Ornstein, A. Lindberg, Å. Ringnér, H. Elmrud, M. Carlsson, A. Runehagen, C. Svanborg and R. Norrby. (1995) "An epidemiologic study of Lyme disease in southern Sweden." *The New England Journal of Medicine,* Vol. 333, No. 20, p. 1319–1324.

Bózsik, B., A. Lakos, J. Budai, L. Telegdy and G. Ambrózy. (1986) "Occurrence of Lyme borreliosis in Hungary." *International Journal of Microbiology and Hygiene (A),* Vol. 263, p. 466–467.

Breitschwerdt, E. (1993) "Tickborne diseases of dogs." *Veterinary Technician,* Vol. 14, No. 5, p. 291–299.

"Bug Off! How to repel biting insects." (1993) *Consumer Reports,* July, p. 451–454.

Burgdorfer, W., A. Barbour, S. Hayes, J. Benach, E. Grunwaldt and J. P. Davis. (1982) "Lyme disease: A tick-borne spirochetosis?" *Science,* Vol. 216, p. 1317–1319.

Burgdorfer, W. (1984) *Ticks: an Ever Increasing Public Health Menace.* New Haven, CT: The Connecticut Agricultural Experiment Station, Bulletin No. 822.

Burgdorfer, W. (1986) "Discovery of the Lyme disease spirochete. A historical review." *International Journal of Microbiology and Hygiene (A),* Vol. 263 p. 7–10.

Burgdorfer, W. (1993) *Status of Tick-borne Disease in the United States.* (speech) Presented at the 6th Annual International Scientific Lyme Disease Conference. Atlantic City, NJ: Lyme Disease Foundation.

Burgess, E. (1986) "Experimental inoculation of dogs with *Borrelia burgdorferi.*" *International Journal of Microbiology and Hygiene (A),* Vol. 263, p. 49–54.

Burgess, E. and L. Patrican. (1987) "Oral infection of *Peromyscus maniculatus* with *Borrelia burgdorferi* and subsequent transmission by *Ixodes dammini.*" *American Journal of Tropical Medicine and Hygiene,* Vol. 36, No. 2, p. 402–407.

Burgess, E., T. Amundson, J. Davis, R. Kaslow and R. Edelman. (1986) "Experimental inoculation of *Peromyscus spp.* with *Borrelia burgdorferi:* Evidence of contract transmission." *American Journal of Tropical Medicine and Hygiene,* Vol. 35, No. 2, p. 355–359.

Burrascano, J. (1997) "Lyme Disease." *Conn's Current Therapies* (Rakel, Ed.), Philadelphia, PA: W. B. Saunders Co., p. 140–143.

Busch, U., C. Hizo-Teufel, R. Boehmer, V. Fingerle, H. Nitschko, B. Wilske and V. Preac-Mursic. (1996) "Three species of *Borrelia burgdorferi* senu lato (*B. burgdorferi* senu stricto, *B. afzelii,* and *B. garinii*) identified from cerebrospinal fluid isolates by pulsed-field gel electrophoresis and PCR." *Journal of Clinical Microbiology,* Vol. 34, No. 5, p. 1072–1078.

Bushmich, S. (1994) "Lyme borreliosis in domestic animals." *Journal of Spirochetal and Tick-borne Diseases,* Vol. 1, No. 1, p. 24–28.

Callister, S., D. Jobe, R. Schell, C. Pavia and S. Lovrich. (1996). "Sensitivity and specificity of the borreliacidal-antibody test during early Lyme disease: A 'gold standard'?" *Clinical and Diagnostic Laboratory Immunology,* Vol. 3, No. 4, p. 399–402.

Cartter, M., P. Mshar, S. Ertel, K. Lawless, S. Brown and J. Hadler. (1996) "Epidemiology of Lyme disease in the Lyme, Connecticut area." *VII International Congress on Lyme Borreliosis.* San Francisco, CA: University of California, Berkeley, p. 74.

"Case Definitions for Public Health Surveillance." (1996) *Centers for Disease Control and Prevention,* Web page (see appendix).

Caution Urged when Using Insect Repellents. (1989) Hartford, CT: State of Connecticut Department of Health Services, August 22.

Chang, Y-F., R. Straubinger, R. Jacobson, J. Kim, T. S. Kim, D. Kim, S. Shin and M. Appel. (1996) "Dissemination of *Borrelia burgdorferi* after experimental infection in dogs." *Journal of Spirochetal and Tick-borne Diseases,* Vol. 3, No. 1, p. 80–86.

Chorba, J. (1995) "Design for deer resistance: Combining flora with fauna." *The green scene,* March.

Chu, H., L. G. Chavez, Jr., B. M. Blumer, R. W. Sebring, T. L. Wasmoen and W. M. Acree. (1992) "Immunogenicity and efficacy study of a commercial *Borrelia burgdorferi* bacterin." *Journal of the American Veterinary Medical Association,* Vol. 201, No. 3, p. 403.

Cimperman, J., F. Strle, V. Maraspin, S. Lotric, E. Ruzk-Sabljic and R. N. Picken.

(1996) "Repeated isolation of *Borrelia burgdorferi* from the cerebrospinal fluid of two patients treated for Lyme neuroborreliosis." *VII International Congress on Lyme Borreliosis.* San Francisco, CA: University of California, Berkeley, p. 181.

Consumer Bulletin—Using Insect Repellents Safely. (1989) Washington, DC: Office of Pesticides and Toxic Substances, United States Environmental Protection Agency, August 15.

Coyle, B., T. Strickland, Y. Liang, C. Peña, R. McCarter and E. Israel. (1996) "The public health impact of Lyme disease in Maryland." *Journal of Infectious Diseases,* Vol. 173, p. 1260–1262.

Coyle, P., Z. Deng, S. Schutzer, A. Belman, J. Benach, L. Krupp and B. Luft. (1993) "Detection of *Borrelia burgdorferi* antigens in cerebrospinal fluid." *Neurology,* Vol. 43, p. 1093–1097.

Coyle, P. (Ed.) (1993) *Lyme Disease.* St. Louis, MO: Mosby Year Book, Inc. Chapters: Belman, A. "Pediatric Lyme disease." p. 210–218; Benach, J. and J. L Coleman. "Overview of spirochetal infections." p. 61–66; Berger, B. "Dermatologic aspects." p. 69–72; Bosler, E. "Tick vectors and hosts." p. 18–26; Burgdorfer, W. "Discovery of *Borrelia burgdorferi.*" p. 3–7; Coleman, J. L. and J. Benach. "Pathogenesis of Lyme disease." p. 179–183; Coyle, P. "Vaccine development." p. 172–178; Duray, P. H. "Histopathology of human borreliosis." p. 49–58; Garcia-Monco, J. C. "European Lyme disease." p. 219–223; Golightly, M. "Antibody assays." p. 115–120; Kaell, A., R. S. Bennett and M. I. Hamburger. "Rheumatic manifestations." p. 73–85; Nowakowski, J. and G. Wormser. "Treatment of early Lyme disease: Infection associated with erythema migrans." p. 149–162; Reik, L. Jr. "Neurologic aspects of North American Lyme disease." p. 101–112; Rosa, P. A. and T. G. Schwan. "Molecular biology of *Borrelia burgdorferi.*" p. 8–17; Schutzer, S. "Immune response and clinical Lyme disease." p. 46–48; Urban, C. and B. Luft. "New antibiotic agents." p. 167–171; Vlay, S. "Cardiac manifestations." p. 86–91.

Craven, B. and G. Miller. "Lyme Disease National Surveillance Case Definition." *Lyme Disease Surveillance Summary,* Vol. 90/1, No.1, p. 1–6.

Curran, K. and D. Fish. (1989) "Increased risk of Lyme disease for cat owners." *New England Journal of Medicine,* Vol. 320, No. 3.

Dattwyler, R., E. Grunwaldt and B. Luft. (1996) "Clarithromycin in treatment of early Lyme disease: A pilot study." *Antimicrobial Agents and Chemotherapy,* Vol. 40, No. 2, p. 468–469.

Dawson, J. and S. Ewing. (1995) "Human ehrlichiosis: A potentially fatal tick-borne disease." *Journal of Spirochetal and Tick-borne Diseases,* Vol. 2, No. 1, p. 19–22.

Dawson, J. and S. Ewing. (1994) "Human granulocytic ehrlichiosis in the upper midwest United States." *Journal of the American Medical Association,* Vol. 272/3, p. 212–218.

Dawson, J., K. Biggie, C. Warner, K. Cookson, S. Jenkins, J. Levine and J. Olson.

(1996) "Polymerase chain reaction evidence of *Ehrlichia chaffeensis,* an etiologic agent of human ehrlichiosis, in dogs from southeast Virginia." *AJVR,* August, p. 1175–1179.

Deinzer, D. (1996). Letter to the CDC stating the results of a study of the two-tier testing (ELISA/western blot). State of New York, Department of Health, April 15.

Deinzer, D., E. Friedlander, D. Suess, G. Birkhead and D. White. (1996) "Lyme disease in NYS: 23,000 cases in ten years." *VII International Congress on Lyme Borreliosis.* San Francisco, CA: University of California, Berkeley, p. 83.

Demaerschalck, I., A. Messaoud, M. DeKesel, B. Hoyois, Y. Lobet, P. Hoet, G. Bigaignon, A. Bollen and E. Godfroid. (1995) "Simultaneous presence of different *Borrelia burgdorferi* genospecies in biological fluids of Lyme disease patients." *Journal of Clinical Microbiology,* Vol. 33, No. 3, p. 602–608.

DeSilva, A. M., S. R. Telford III, L. R. Brunet, S. W. Barthold and E. Fikrig. (1996) "*Borrelia burgdorferi* OspA is an arthropod-specific transmission-blocking Lyme disease vaccine." *Journal of Experimental Medicine,* Vol. 183, No. 1, p. 271–275.

Donta, S. T. (1995) "Lyme disease: A clinical challenge." *Journal of Spirochetal and Tick-borne Diseases,* Vol. 2, No. 3, p. 50–51.

Dorward, D. and C. Garon. (1994) "Immune capture and cultivation of *Borrelia burgdorferi.*" *Journal of Spirochetal and Tick-borne Diseases,* Vol. 1, No. 4, p. 85–89.

Drummond, R. (1990) *Ticks and What You Can Do About Them.* Berkeley, CA: Wilderness Press.

Dumler, J. S. (1996) "Is human granulocytic ehrlichiosis (HGE) another Lyme disease?" (speech) *Lyme Disease Foundation 9th Annual International Scientific Conference on Lyme Borreliosis and other Tick-borne Disorders.* Hartford, CT: Lyme Disease Foundation.

Dunne, M., B. K. al-Ramadi, S. Barthold, R. A. Flavell and E. Fikrig. (1995) "Oral vaccination with an attenuated *Salmonella typhimurium* strain expressing *Borrelia burgdorferi* OspA prevents murine Lyme borreliosis." *Infection and Immunity,* Vol. 63, p. 1611–1614.

"Dursban and birth defects." (1996) *ABDC NEWS* (Association of Birth Defects in Children Newsletter), Vol. 26, Issue 2.

Evans, S. R., G. W. Korch and M. A. Lawson. (1990) "Comparative field evaluation of permethrin and deet-treated military uniforms for personal protection against ticks (Acari: Ixodidae)." *Journal of Medical Entomology,* Vol. 27, p. 829–834.

Fallon, B. and J. Nields. (1994) "Lyme disease: A neuropsychiatric illness." *American Journal of Psychiatry,* Vol. 151, No. 11, p. 1571–1582.

Fallon, B., H. Bird, C. Hoven, D. Cameron, M. Liebowitz and D. Shaffer. (1994) "Psychiatric aspects of Lyme disease in children and adolescents: A community epidemiologic study in Westchester, N.Y." *Journal of Spirochetal and Tick-borne Diseases,* Vol. 1, No. 4, p. 98–101.

Fallon, B., J. Nields, J. Burrascano, K. Liegner, D. DelBene and M. Liebowitz. (1992)

"The neuropsychiatric manifestations of Lyme borreliosis." *Psychiatric Quarterly,* Vol. 63, No.1, p. 95–117.

Farley, J., S. Banerjee and D. O'Hanlon. (1990) "Lyme disease in BC: A review." *British Columbia Medical Journal,* Vol. 32, No. 10, p. 432–435.

Farrell, G. M. and E. H. Marth. (1991) "*Borrelia burgdorferi:* Another cause of foodborne illness?" *International Journal of Food Microbiology,* Vol. 14 (3–4), p. 247–260.

Feder, H. and M. Hunt. (1995) "Pitfalls in the diagnosis and treatment of Lyme disease in children." *Journal of the American Medical Association,* July, p. 66–68.

Feir, D. and C. Reppell. (1990) "*Borrelia burgdorferi* in Missouri." *The Missouri Academy of Science,* Occasional Paper, No. 8, p. 12–18.

Feir, D., C. Repell-Santanello, B. W. Li, C. S. Xie, E. Masters, R. Marconi and G. Weil. (1994) "Evidence supporting the presence of *Borrelia burgdorferi* in Missouri." *American Journal of Tropical Medicine and Hygiene,* Vol 51, No. 4, p. 475–482.

Felsenfeld, O. (1971) *Borrelia: Strains, vectors, human and animal borreliosis*. St. Louis, MO: Warren H. Green, Inc.

Fikrig, E., S. W. Barthold, F. S. Kantor and R. A. Flavell. "Protection of mice against the Lyme disease agent by immunizing with recombinant OspA." *Science,* Vol. 250, p. 553–555.

Fingerle, V., H. Bergmeister, G. Liegl, E. Vanek and B. Wilske. (1994) "Prevalence of *Borrelia burgdorferi* sensu lato in *Ixodes ricinus* in southern Germany: *Borrelia* infection of *Ixodes ricinus*." *Journal of Spirochetal and Tick-borne Diseases,* Vol. 1, No. 2, p. 41–45.

Fister, R. and R. Tilton. Personal communication regarding unpublished work, October, 1996.

Garbe, C., H. Stein, D. Dienemann and C. Orfanos. (1991) "*Borrelia burgdorferi*-associated cutaneous B-cell lymphoma: Clinical and immunohistologic characterization of four cases." *Journal of the American Academy of Dermatology,* Vol. 24, p. 584–590.

Garcia-Monco, J. C. and J. Benach. (1994) "Approaches to the laboratory diagnosis of Lyme neuroborreliosis." *Clinical Immunology Newsletter,* Vol. 14, No. 12 p. 153–156.

Gardner, T. (1995) "Lyme disease." *Infectious Diseases of the Fetus and Newborn.* New York, NY: Remington-Saunders, p. 447–528.

Ginsberg, H. (1993) *Ecology and Environmental Management of Lyme Disease*. New Brunswick, NJ: Rutgers University Press.

Goddard, Capt. J. (1988) "Dispelling the myths about tick removal." *USAF Medical Service Digest,* Spring, p. 16.

Gorelova, N. B., D. Postic, E. I. Korenberg, G. Baranton, E. Bellengen and Y. V. Kovalevskii. (1996) "*Borrelia* genospecies mixtures in ticks and small mammals from natural foci." (France/Russia) *VII International Congress on Lyme Borreliosis.* San Francisco, CA: University of California, Berkeley, p. 20.

Guttman, D., P. Wang, I.-N. Wang, E. Bosler, B. Luft and D. Dykhuizen. (1996) "Multiple infections of *Ixodes scapularis* ticks by *Borrelia burgdorferi* as revealed by single-strand conformation polymorphism analysis." *Journal of Clinical Microbiology,* Vol. 34, No. 3, p. 652–656.

Harris, N. S. and B. G. Stephens. (1995) "Detection of *Borrelia burgdorferi* antigen in urine from patients with Lyme borreliosis." *Journal of Spirochetal and Tick-borne Diseases,"* Vol. 2, No. 2, p. 37–41.

Häupl, T., G. Hahn, M. Rittig, A. Krause, C. Schoerner, U. Schönherr, J. Kalden and G. Burmester. (1993) "Persistence of *Borrelia burgdorferi* in ligamentous tissue from a patient with chronic Lyme borreliosis." *Arthritis and Rheumatism,* Vol. 36, No. 11, p. 1621–1626.

Heir, G. and L. Fein. (1996) "Lyme Disease: Considerations for dentistry." *Journal of Orofacial Pain,* Vol. 10, No. 1, p. 74–86.

Hercogová, J., D. Hulínská, J. Zivny and D. Janovská. (1996) "Erythema migrans during pregnancy: Study of 35 women." (1996) *VII International Congress on Lyme Borreliosis.* San Francisco, CA: University of California, Berkeley, p. 172.

Hilton, E., J. Devoti and S. Sood. (1996) "Recommendation to include OspA and OspB in the new immunoblotting criteria for serodiagnosis of Lyme disease." *Journal of Clinical Microbiology,* Vol. 34, No. 6, p. 1353–1354.

Hoskins, J. and E. Cup. (1988) "Ticks of veterinary importance. Part I. The Ixodidae family: Identification, behavior, and associated diseases." *Compendium Small Animal,* Vol. 10, No. 5, p. 565–580.

Hudson, B., D. Barry, D. Shafren, M. Wills, S. Caves and V. Lennox. (1994) "Does Lyme borreliosis exist in Australia?" *Journal of Spirochetal and Tick-borne Diseases,* Vol. 1, No. 2, p. 46–51.

Húlinská, D., J. Basta, R. Murgia and M. Cinco. (1995) "Intracellular morphological events observed by electron microscopy on neutrophil phagocytosis of *Borrelia garinii.*" *Journal of Spirochetal and Tick-borne Diseases,* Vol. 2, No. 4, p. 82–86.

"Insect repellants." (1989) *Medical Letter on Drugs and Therapeutics,* Vol. 31, Issue 792, p. 1–3.

Johnson, R. C., G. P. Schmid, F. W. Hyde, A. G. Steigerwalt and D. J. Brenner. (1984) "*Borrelia burgdorferi* sp. nov: Etiologic agent of Lyme disease." *International Journal of Systematic Bacteriology,* Vol. 34, p. 496–497.

Johnson, R., C. Kodner, M. Russell and D. Girard. (1990) "In-vitro and in-vivo susceptibility of *Borrelia burgdorferi* to Azithromycin." *Journal of Antimicrobial Chemotherapy,* Vol. 25, Supplement A, p. 33–38.

Johnson, R., C. Kodner, P. Jurkovich and J. Collins. (1990) "Comparative in vitro and in vivo susceptibilities of the Lyme disease spirochete *Borrelia burgdorferi* to Cefuroxime and other antimicrobial agents." *Antimicrobial Agents and Chemotherapy,* Vol. 34, No. 11, p. 2133–2136.

Johnson, S., B. Swaminathan, P. Moore, C. Broome and M. Parvin. (1990) "*Borrelia burgdorferi*: Survival in experimentally infected human blood processed for transmission." *Journal of Infectious Diseases,* Vol. 162, p. 557–559.

Katzung, B. (1995) *Basic and Clinical Pharmacology.* Norwalk, CT: Appleton and Lange.

Katzung, B. and A. Trevor. (1995) *Examination and Board Review Pharmacology.* Norwalk, CT: Appleton and Lange.

Kazarian, D. (1992) *Pharmacology of Antibiotic Choices.* Clearwater, FL: Infuserve America.

Keller, D., F. Koster, D. Marks, P. Hosbach, L. Erdile and J. Mays. (1994) "Safety and Immunogenicity of a recombinant outer surface protein A Lyme vaccine." *Journal of the American Medical Association,* Vol. 271, No. 22, p. 1764–1768.

Keszler, K. and R. Tilton. (1995) "Persistent PCR positivity in a patient being treated for Lyme disease." *Journal of Spirochetal and Tick-borne Diseases*, Vol. 2, No. 3, p. 57–58.

Klempner M., R. Noring, M. Peacocke, K. Georgilis, C. Braden and R. Rogers (1992) "Invasion of human skin fibroblasts by the Lyme disease spirochete *Borrelia burgdorferi.*" *Program Abstracts. 5th International Conference on Lyme Borreliosis.*

Klempner, M., R. Noring and R. Rogers. (1993) "Invasion of human skin fibroblasts by the Lyme disease spirochete, *Borrelia burgdorferi.*" *Journal of Infectious Diseases,* Vol. 167, p. 1074–1081.

Krause, P., S. Telford III, A. Spielman, V. Sikand, R. Ryan, D. Christianson, G. Burke, P. Brassard, R. Pollack, J. Peck and D. Persing. (1996) "Concurrent Lyme disease and babesiosis. Evidence for increased severity and duration of illness." *Journal of the American Medical Association,* Vol. 275, No. 21, p. 1657–1660.

Kütting, B., G. Bonsmann, D. Metze, T. Luger and L. Cerroni. (1997) "*Borrelia burgdorferi*-associated primary cutaneous B-cell lymphomas with *Borrelia* infection proved by cultivation from lesional skin." *Journal of the American Academy of Dermatology* (in press).

Lane, R. (1994) "Ticks of California and their public health significance." *Journal of Spirochetal and Tick-borne Diseases,* Vol. 1, No. 3, p. 74–76.

Langermann, S., S. Paloazynskl, A. Sadziene, K. Stover and S. Koenig. (1994) "Systemic and mucosal immunity induced by B.C.G. vector expressing outer-surface protein A of *Borrelia burgdorferi.*" *Nature,* Vol. 378, p.552–556.

Langreth, R. (1995) "Obscure tick-borne disease threatens golfers and others who brave the woods." *Wall Street Journal,* August 17.

Lavoie, P. E., A. J. Wilson and D. L. Tuffanelli. (1986) "Acrodermatitis chronica atrophicans with antecedent Lyme disease in a Californian." *Zentralbl Bakteriol Hyg A,* Vol. 263, p. 262–265.

Lawrence, C., R. Lipton, F. Lowy and P. Coyle. (1995) "Seronegative chronic relapsing neuroborreliosis." *European Neurology,* Vol. 35, p. 113–117.

Lesser, R., E. Kornmehl, A. Pachner, J. Kattah, T. Hedges III, N. Newman, P. Ecker and M. Glassman. (1990) "Neuro-Ophthalmologic manifestations of Lyme disease." *Ophthalmology,* Vol. 97, No. 6, p. 699–706.

Levy, S., D. Dombach, S. Barthold and T. Wasmoen. (1993) "Canine Lyme borreliosis." *The Compendium,* Vol. 15, No. 6, p. 833–846.

Li, Y., Z. Zhang, X. Liu, B. Zhou and G. Wang. (1996) "Lyme neuroborreliosis associated with Guillian-Barré syndrome." *VII International Congress on Lyme Borreliosis.* San Francisco, CA: University of California, Berkeley, p. 123.

Liegner, K., J. Shapiro, D. Ramsay, A. Halperin, W. Hogrefe and L. Kong. (1993) "Recurrent erythema migrans despite extended antibiotic treatment with minocycline in a patient with persisting *Borrelia burgdorferi* infection." *Journal of the American Academy of Dermatology,* February, p. 312–314.

Logigian, E., R. Kaplan and A. Steere. (1990) "Chronic neurologic manifestations of Lyme disease." *New England Journal of Medicine,* November, p. 1438–1444.

Luft, B., R. Dattwyler, R. Johnson, S. Luger, E. Bosler, D. Rahn, E. Masters, E. Grunwaldt and S. Gadgil. (1996) "Azithromycin compared with amoxicillin in the treatment of erythema migrans." *Annals of Internal Medicine,* Vol. 124, No. 9, p. 785–791.

Luger, S., P. Paparone, G. Wormser, R. Nadelman, E. Grunwaldt, G. Gomez, M. Wisniewski and J. Collins. (1995) "Comparison of Cefuroxime Axetil and doxycycline in treatment of patients with early Lyme disease associated with erythema migrans." *Antimicrobial Agents and Chemotherapy,* Vol. 39, No. 3 p. 661–667.

"Lyme disease." (1990) *Morbidity and Mortality Weekly Report,* October, p. 19–21.

"Lyme disease—United States, 1995." (1996) *Morbidity and Mortality Weekly Report,* June, p. 481–484.

Ma, Y., A. Sturrock and J. Weis. (1991) "Intracellular localization of *Borrelia burgdorferi* within human endothelial cells." *Infection and Immunity,* Vol. 59, p. 671–678.

MacDonald, A., J. Benach and W. Burgdorfer. (1987) "Stillbirth following maternal Lyme disease." *New York State Journal of Medicine,* Vol. 87, p. 615–616.

Magnarelli L., K. Stafford III and V. Bladen. (1992) "*Borrelia burgdorferi* in *Ixodes dammini* (Acari: Ixodidae) feeding on birds in Lyme, Connecticut, U.S.A." *Canadian Journal of Zoology,* Vol. 70, p. 2322–2325.

Mandell, G., G. Douglas, Jr. and J. Bennett. (Ed.) (1990). *Principles and Practices of Infectious Diseases.* New York, NY: Churchill Livingstone Inc. Chapters: Boyce, J. "Francisella tularensis (tularemia)." pp. 1742–1748; Donowitz, G. and G. Mandell. "Cephalosporins." p. 246–257; Farrar, E. "Leptospira species (Leptospirosis)." p. 1813–1816; Gelfand, J. "*Babesia.*" p. 2119–2122; Johnson, Jr., W. "Borrelia species (relapsing fever)." p. 1816–1819; Lietman, P. "Pharmacokinetics of antimicrobial agents." p. 228–230; Mayer, K., S. Opal and A. Medeiros. "Mechanisms of antibiotic resistance." p. 218–228; Menegus, M. and R. Douglas, Jr. "Viruses, rickettsiae, chlamydiae, and mycoplasmas." p. 193–205; Moellering, Jr., R. "Principles of anti-infective therapy." p. 206–218; Monath, T. "Colorado tick fever." p. 1233–1234; Monath, T. "Reovirus and orbivirus." p. 1234–1235; Neu, H. "Penicillins." p. 230–246; Raoult, D. and D. Walker. "*Rickettsia rickettsii* and other spotted fever group rickettsiae (Rocky Mountain spotted fever and other spotted fevers)." p. 1465–1471; Saah, A. "Introduction to rickettsiosis." p. 1463–1465; Saah, A. "Ehrlichia species (human ehrlichiosis)."

p. 1482–1483; Standiford, H. "Tetracyclines and chloramphenicol." p. 284–295; Steigbigel, N. "Erythromycin, lincomycin and clindamycin." p. 308–317; Steere, A. "*Borrelia burgdorferi* (Lyme disease, Lyme borreliosis)." p. 1819–1827; Washington II, J. "Bacteria, fungi, and parasites." p. 160–193; Weber, D., W. Gammon and M. Cohen. "The acutely ill patient with fever and rash." p. 479–489; Wilson, B. and P. Weary. "Ticks (including tick paralysis)." p. 2168–2170.

Marshall, W. F., S. R. Telford III, P. N. Rys et al. (1994) "Detection of *Borrelia burgdorferi* DNA in museum specimens of *Peromyscus leucopus*." *Journal of Infectious Diseases,* Vol. 170, p. 1027–1032.

Mast, W. E. and W. M. Burrows. (1976) "Erythema chronicum migrans in the United States." *Journal of the American Medical Association*, Vol. 236, p. 859–860.

Masters, E. and D. Donnell. "Lyme and/or Lyme-like disease in Missouri." (1995) *Missouri Medicine,* Vol. 92, No. 7, p. 346–353.

Mather, T., D. Fish and R. Coughlin. (1994) "Competence of dogs as reservoirs for Lyme disease spirochetes (*Borrelia burgdorferi*)." *Journal of the American Veterinary Medical Association,* Vol. 205, No. 2, p. 186–188.

Matuschka, F., A. Ohlenbusch, H. Eiffert, D. Richter and A. Spielman. (1995) "Antiquity of the Lyme-disease spirochete in Europe." *The Lancet,* Nov. 18. p. 1367.

May, E. and B. Jabbari. (1990) "Stroke in neuroborreliosis." *Stroke,* Vol. 21, No. 8, p. 1232–1235.

"Medimmune reports data on new Lyme disease vaccine candidate at international Lyme meeting." (1996) *Medimmune Press Release,* June.

Middleton, D. (1994) "Tick-borne infections." *Postgraduate Medicine,* Vol. 95, No. 5, p. 131–139.

Miller, L. (1996) Personal correspondence on the symptoms of Lyme in animals, antibody titers/their interpretations and other means of animal diagnosis of Lyme disease.

Miyamoto, K., Y. Sato and F. Sato. (1996) "Isolation of *Borrelia burgdorferi* sensu lato from migratory birds, *Turdus chrysolaus,* at Nemuro, Hokkaido." *VII International Congress on Lyme Borreliosis.* San Francisco, CA: University of California, Berkeley, p. 79.

Montgomery, R., M. Nathanson and S. Malawista. (1991) "Intracellular fate of *Borrelia burgdorferi* in mouse macrophages." *Arthritis and Rheumatism,* Vol. 34, p. S50.

Moorhouse, D. (1969) "The attachment of some *Ixodid* ticks to their natural hosts." *Proceedings of the 2nd International Congress of Acarology* (1967), p. 319–327.

Mount, G. A. and E. L. Snoddy. (1983) "Pressurized sprays of permethrin and deet on clothing for personal protection against the lone star tick and the American dog tick. (Acari: Ixodidae)." *Journal of Economic Entomology,* Vol. 76, No. 3, p. 529–531.

Muellegger, R., G. Brunner-Koehler, H. P. Soyer, S. Hoedl, H. Kerl and M. Millner. (1996) "Dermatoborreliosis (DB) in children (CH) in a European area endemic for Lyme borreliosis (LB) (Styria/Austria) between 1993 and 1995." *VII International Congress on Lyme Borreliosis.* San Francisco, CA: University of California, Berkeley, p. 107.

Murray, P., E. Baron, M. Pfaller, F. Tenover and R. Yolken. (Ed.) (1995) *Manual of Clinical Microbiology*. Washington, DC: American Society for Microbiology. Chapters: Herrmann, J. "Immunoassays for the diagnosis of infectious diseases." p. 110–122; Baron, E. J., A. Weissfeld, P. Fuselier and D. Brenner. "Classification and identification of bacteria." p. 249–264; Stewart, S. "Francisella." p. 545–548; Schwan, T., W. Burgdorfer and P. Rosa. "Borrelia." p. 626–635; Olson, J. and J. McDade. "Rickettsia and Coxiella." p. 678–685; Olson, J. and J. Dawson. "Ehrlichia." p. 686–689; Yao, J. and R. Moellering, Jr. "Antibacterial agents." p. 1281–1307.

Nakoa, M. and K. Miyomoto. (1995) "Mixed infection of different *Borrelia* species among *Apodemus speciosus* mice in Hokkaido, Japan." *Journal of Clinical Microbiology,* February, p. 490–492.

Needham, G. (1985) "Evaluation of five popular methods for tick removal." *Pediatrics,* Vol. 75, No. 6, p. 997–1002.

New Medicines in Development for Infectious Diseases. (1996) Washington, DC: Pharmaceutical Research and Manufacturers of America.

New Yorkers Urged to use Lower Concentration deet Products. (1991) Albany, NY: NYS Department of Health Press Release, May 22, p. 1–2.

O'Connell, J. Robertson, P. Stewart, J. White and E. Guy. (1996) "Lyme borreliosis in a UK endemic area: Long-term follow-up, changing epidemiology and effects of education." *VII International Congress on Lyme Borreliosis.* San Francisco, CA: University of California, Berkeley, p. 64.

Oliver, Jr., J. H., M. R. Owslen, H. J. Hutcheson, A. M. James, J. C. Chen, W. S. Irby, E. M. Dotson and D. K. McLain. (1993) "Conspecificity of the ticks *Ixodes scapularis* and *I. dammini* (Acari:Ixodidae)." *Journal of Medical Entomology,* Vol. 30, p. 54–63.

Olsen, B., D. Duffy, T. G. Jaenson, A. Gylfe, J. Bonnedahl and S. Bergstrom. (1995) "Transhemispheric exchange of Lyme disease spirochetes by seabirds." *Journal of Clinical Microbiology,* Vol. 33, No. 12, p. 3270–3274.

Padilla, M., S. Callister, R. Schell, G. Bryant, D. Jobe, S. Lovrich, B. DuChateau and J. Jensen. (1996) "Characterization of the protective borreliacidal antibody response in humans and hamsters after vaccination with a *Borrelia burgdorferi* outer surface protein A vaccine." *Journal of Infectious Diseases*, Vol. 174, p. 739–746.

Paparone, P. and P. Paparone. (1995) "Lyme disease in the elderly." *Journal of Spirochetal and Tick-borne Diseases,* Vol. 2, No. 1, p. 14–18.

Pape, J. (1995) "1994 Tick-borne disease summary." *State of Colorado Communicable Disease Epidemiology,* April 10, p. 1–2.

Parker, J. and K. White. (1992) "Lyme borreliosis in cattle and horses: A review of the literature." *Cornell Vet,* Vol. 82, No. 3, p. 253–274.

Parkinson, D., B. Luft, A. Devito and R. Dattwyler. (1996) "Prevalence of Lyme disease in outside workers in Long Island." *VII International Congress on Lyme Borreliosis.* San Francisco, CA: University of California, Berkeley, p. 108.

Patmas, M. and C. Remorca. (1994) "Disseminated Lyme disease after short-duration tick-bite." *Journal of Spirochetal and Tick-borne Diseases,* Vol. 1, No. 3, p. 77–78.

Pavia, C. (1994) "Overview of pathogenic spirochetes." *Journal of Spirochetal and Tick-borne Diseases,* Vol. 1, No. 1, p. 3–11.

Pfister, H.-W., K. M. Einhäupl, V. Preac-Mursic, B. Wilske and G. Schierz. (1984) "The spirochetal etiology of lymphocytic meningoradiculitis of Bannwarth (Bannwarth's syndrome)." *Journal of Neurology,* Vol. 231, p. 141–144.

Philipp, M., Y. Lobet, R. Bohm, M. Conway, V. Dennis, P. Desmons, Y. Gu, P. Hauser, R. Lowrie and D. Roberts. "Safety and immunogenicity of recombinant outer surface protein A (OspA) vaccine formulations in the rhesus monkey." *Journal of Spirochetal and Tick-borne Diseases,* Vol. 3, No. 1, p. 67–79.

Pichon, B., E. Godfriod, B. Hoyois, A. Bollen, F. Rodhain and C. Perez-Eid. (1995) "Simultaneous infection of *Ixodes ricinus* nymphs by two *Borrelia burgdorferi* sensu lato species: Possible implications for clinical manifestations." *Emerging Infectious Diseases,* Vol. 1, No. 3, p. 89.

Pietrucha, D. (1996) *Neurologic manifestations of Lyme disease in children.* Neptune, NJ: www.Lymenet.org.

Post, J. (1990) *A Review of Laboratory Diagnostic Tests for Lyme Borreliosis.* Storrs, CT: University of Connecticut.

Preac-Mursic, V., E. Patsouris, B. Wilske, S. Reinhardt, B. Gross and P. Mehraein. (1990) "Persistence of *Borrelia burgdorferi* and histopathological alterations in experimentally infected animals. A comparison with histopathological findings in human Lyme disease." *Infection,* Vol. 18, No. 6, p. 332–341.

Preac-Mursic, V., K. Weber, H.-W. Pfister, B. Wilske, B. Gross, A. Baumann and J. Prokop. (1989) "Survival of *Borrelia burgdorferi* in antibiotically treated patients with Lyme borreliosis." *Infection,* Vol. 17, No. 6, p. 355–359.

Preac-Mursic, V., W. Marget, U. Busch, P. Rigler and S. Hagl. (1996) "Kill kinetics of *Borrelia burgdorferi* and bacterial findings in relation to the treatment of Lyme borreliosis." *Infection,* Vol. 24, No. 1, p. 9–16.

Preac-Mursic, V., B. Wilske and G. Scherz. (1986) "European *Borrelia burgdorferi* isolated from humans and ticks. Culture conditions and antibiotic susceptibility." *Zentral Bakteriol hyg (A),* Vol. 263, p. 112–118.

Prince, G. and S. N. Banerjee. (1995) "Lyme arthritis in British Columbia." *Journal of Spirochetal and Tick-borne Diseases,* Vol. 2, No. 3, p. 52–54.

Rawlings, J. (1995) "Tick-borne disease in Texas, 1994." *Disease Prevention News,* Vol. 55, No. 11, p. 1–8.

Rawlings, J. and G. Teltow. (1994) "Prevalence of *Borrelia* (Spirochaetaceae) spirochetes in Texas ticks." *Journal of Medical Entomology,* Vol. 31, No. 2, p. 297–301.

"Recommendations for test performance and interpretation from the second national conference on serologic diagnosis of Lyme disease." (1995) *Morbidity and Mortality Weekly Report,* Vol. 44, No. 31, p. 1–3.

Reik, L. (1992) *Lyme Disease and the Nervous System.* New York, NY: Thieme (as cited in Fallon, B. and J. Nields, "Lyme disease: A neuropsychiatric illness." (1994) *American Journal of Psychiatry,* p. 1571–1582.

Rich, S. M., D. A. Caporale, S. R. Telford III, T. D. Kocher, D. L. Hartl and A. Spielman. (1995) "Distribution of the *Ixodes ricinus*-like ticks of eastern North America." *Proceedings of the National Academy of Science,* Vol. 92, p. 6284–6288.

Rucker, D. (1984) *Impact: TVA—Natural Resources and the Environment,* (Tennessee Valley Authority newsletter), Vol. 7, No. 4.

Schlesinger, P., P. Duray, B. Burke, A. Streere and T. Stillman. (1985) "Maternal-fetal transmission of the Lyme disease spirochete *Borrelia burgdorferi.*" *Annals of Internal Medicine,* Vol. 103, p. 67–68.

Schmidt, B., E. Aberer, C. Stockenhuber, Ch. Wagner, H. Klade, F. Breir and A. Luger. (1995) "Detection of *Borrelia burgdorferi*-DNA in urine from patients with Lyme borreliosis." *Journal of Spirochetal and Tick-borne Diseases,* Vol. 2, No. 4, p. 76–81.

Schmidt, B., E. Aberer, C. Stockenhuber, H. Klade, F. Breir and A. Luger. (1995) "Detection of *Borrelia burgdorferi* DNA by polymerace chain reaction in the urine and breast milk of patients with Lyme borreliosis." *Diagn Microbiol Infect Dis,* Vol. 21, p. 121–128.

Schreck, C., D. Fish and T. P. McGovern. (1995) "Activity of repellents applied to skin for protection against *Amblyomma americanum* and *Ixodes scapularis* ticks (Acari: Ixodidae)." *Journal of the American Mosquito Control Association,* Vol. 11, No. 1, p. 136–140.

Schreck, C. E., E. L. Snoddy and A. Spielman. (1986) "Pressurized sprays of permethrin or deet on military clothing for personal protection against *Ixodes dammini* (Acari: Ixodidae)." *Journal of Medical Entomology,* Vol. 23, No. 4 p. 396–399.

Schulze, T., G. Bowen, E. Bosler, M. Lakat, W. Parkin, R. Altman, B. Ormiston and J. Shisler. (1984) "*Amblyomma americanum*: a potential vector of Lyme disease in New Jersey." *Science,* Vol. 224, p. 601–603.

Schulze, T., L. Vasvary and R. Jordan. (1993) *Lyme Disease: Assessment and Management of Vector Tick Populations in New Jersey.* New Brunswick, NJ: Rutgers Cooperative Extension.

Schutzer, S. (1992) *Lyme Disease Molecular and Immunologic Approaches.* Plainview, NY: Cold Spring Harbor Laboratory Press. Chapters: Benach, J. and J. Garcia-Monco. "Aspects of the pathogenesis of neuroborreliosis." p. 1–10; Duray, P. H. "Target organs of *Borrelia burgdorferi* infections: Functional responses and histology." p. 11–30; Fikrig, E., S. Barthold, J. Sears, S. Telford III, A. Spielman, F. Kantor and R. Flavell. "A recombinant vaccine for Lyme disease." p. 263–282; Girons, I. S. and B. E. Davidson. "Genome organization of *Borrelia burgdorferi.*" p. 111–118; Lukehart, S. A. and C. Marra. "A comparison of syphilis and Lyme disease: Central nervous system involvement." p. 59–77.

Schwan, T., K. Gage and J. Hinnebusch. (1995) "Analysis of relapsing fever spirochetes from the western United States." *Journal of Spirochetal and Tick-borne Diseases,* Vol. 2, No.1, p. 3–8.

Schwan, T., J. Piesman, W. T. Golde, M. C. Dolan and P. Rosa. (1995) "Induction of an outer surface protein on *Borrelia burgdorferi* during tick feeding." *Proceedings of the National Academy of Science,* Vol. 92, p. 2909–2913.

Scrimenti, R. (1970) "Erythema chronicum migrans." *Archives of Dermatology,* Vol. 102, p. 104–105.

Scrimenti, R. (1995) "Acrodermatitis chronica atrophicans: Historical and clinical overview." *Journal of Spirochetal and Tick-borne Diseases,* Vol. 2, No. 4, p. 97–100.

"Seizures temporally associated with use of deet insect repellent—New York and Connecticut." (1989) *Morbidity and Mortality Weekly Report,* Vol. 38, No. 39, p. 678–680.

Shih, C., A. Spielman and S. R. Telford III. (1995) "Short report: Mode of action of protective immunity to Lyme disease spirochetes." *American Journal of Tropical Medicine and Hygiene,* Vol. 52, No. 1, p. 72–74.

Smoak, B., J. McClain, J. Brundage, L. Broadhurst, D. Kelly, G. Dasch and R. Miller. (1996) "An outbreak of spotted fever rickettsiosis in U.S. army troops deployed to Botswana." *Emerging Infectious Diseases,* Vol. 2, No. 3, p. 217–221.

Spencer-Molloy, F. (1995) "Trial of Lyme vaccine begins." *The Hartford Courant,* January 25.

Spielman, A., C. Clifford, J. Piesman and M. Corwin. (1979) "Human babesiois on Nantucket Island, USA: Description of the vector, *Ixodes dammini,* n. sp. (Acarina: Ixodidae)." *Journal of Medical Entomology,* Vol. 15, p. 218–234.

Stafford III, K. (1993) "Reduce abundance of *Ixodes scapularis* (Acari: Ixodidae) with exclusion of deer by electric fencing." *Journal of Medical Entomology,* Vol. 30, No. 6, p. 986–996.

Stafford III, K. (1994) "Several environmental modifications reduce risk of tick bite." *Frontiers of Plant Science,* Vol. 46, No. 2, p. 5–6.

Stanek, G., W. Kristoferitsch, M. Pletschette, A. G. Barbour and H. Flamm. (Ed.) (1989) *Lyme Borreliosis II.* New York, NY: Gustav Fischer Verlag. Chapters: Åsbrink, E., A. Hovmark and I. Olsson. "Lymphodenosis benigna cutis solitaria—Borrelia lymphocytoma in Sweden." p. 156–163; Barbour, A. G. "Classification of *Borrelia burgdorferi* on the basis of plasmids profiles." p. 1–7; Barontini, F., F. Gori and S. Maurri. "A case of Lyme borreliosis presenting as Bannwarth's syndrome with fatal outcome due to a disseminated intravascular coagulation." p. 250–255; Burgdorfer, W. "*Borrelia burgdorferi*: Its relationship to tick vector." p. 8–13; Fahrer, H., S. van der Linden, M.-J. Sauvain, L. Gern, E. Zhioua and A. Aeschlimann. "A positive 'Lyme serology'—What does it mean clinically? Preliminary results of a Swiss prospective study." p. 329–333; Gern, L., E. Frossard, A. Walter and A. Aeschlimann. "Presence of antibodies against *Borrelia burgdorferi* in a population of the Swiss Plateau." p. 321–328; Hanefeld, F., H.-J. Christen, N. Bartlau, K. Wassermann and R. Thomssen. "Lyme borreliosis in children." p. 192–202; Kristoferitsch,

W., E. Sluga, M. Graf, H. Partsch, E. Aberer, R. Neumann and G. Stanek. "Acrodermatitis chronica Atrophicans-associated Neuropathies due to Vasculitis?" p. 126–131; Lakos, A. "Lyme borreliosis in Hungary—The first three years." p. 55–59; Omasits, M. and M. Brainin. "Neuroborreliosis: Finding on patients with either early or late treatment." p. 280; Paul, H., R. Ackermann and H.-J. Gerth. "Infection and manifestation rate of European Lyme borreliosis in humans." p. 44–49; Pejcoch, M., Z. Kralikova, P. Strnad and G. Stanek. "Prevelance to antibodies to *Borrelia burgdorferi* in forestry workers in South Moravia." p. 317–320; Rockstroh, T., H. Mochmann and G. Stanek. "Lyme borreliosis by contact infection, a case report." p. 40–41; Stanek, G. and J. Simeoni. "Are pigeons' ticks transmitters of *Borrelia burgdorferi* to humans? A preliminary report." p. 42–43; Stanek, G., A. Prinz, G. Wewalka, A. Hirschl and Kebela-Ilunga. "Lyme borreliosis in Central Africa." p. 77–81; Strle, F., A Pejovnik-Pustinek, G. Stanek, D. Pleterski and R. Rakar. "Lyme borreliosis in Slovenia in 1986." p. 50–54; Wassmann, K., M. Borg-Von Zepelin, O. Zimmermann, M. Stadler, H. Eiffert and R. Thomssen. "Determination of immunoglobulin M antibody against *Borrelia burgdorferi* to differentiate between acute and past infection." p. 281–289; Weber, K. "Clinical differences between European and North-American Lyme borreliosis—review." p. 146–155; Weber, K. and R. Thurmayr. "Oral penicillin versus minocycline for the treatment of early Lyme borreliosis." p. 263–268; Wilske, B., V. Preac-Mursic, G. Schierz, G. Liegl and W. Gueye. "Detection of IgM- and IgG-antibodies to *Borrelia burgdorferi* using different strains as antigen." p. 299–309.

Steere, A. and S. Malawista. (1979) "Cases of Lyme disease in the United States: Locations correlated with distribution of *Ixodes dammini*." *Annals of Internal Medicine,* Vol. 91, p. 730–733.

Steere, A., A. Pachner and S. Malawista. (1983) "Neurologic abnormalities of Lyme disease: Successful treatment with high-dose intravenous penicillin." *Annals of Internal Medicine,* Vol. 99, No. 6, p. 767–772.

Steere, A. C., et al (1983) "The spirochetal etiology of Lyme disease." *New England Journal of Medicine,* Vol. 308, p. 733–740.

Steere, A., S. Malawista, J. Hardin, S. Ruddy, P. Askenase and W. Andiman. (1977a) "Erythema chronicum migrans and Lyme arthritis: The enlarging clinical spectrum." *Annals of Internal Medicine,* Vol. 86, No. 6, p. 685–698.

Steere, A., S. Malawista, D. Snydman, R. Shope, W. Andiman, M. Ross and F. Steele. (1977b) "Lyme arthritis. An epidemic of oligoarticular arthritis in children and adults in three Connecticut communities." *Arthritis and Rheumatism,* Vol. 20, p. 7–17.

Steinberg, S., T. Strickland, C. Peña and E. Israel. (1996) "Lyme disease surveillance in Maryland, 1992." *Annals of Epidemiology,* Vol. 6, No. 1, p. 24–29.

Strickland, T., I. Caisley, M. Woubeshet and E. Israel. (1994) "Antibiotic therapy for Lyme disease in Maryland." *Public Health Reports,* Vol. 109, No. 6, p. 745–749.

"Summary of Notifiable Diseases, United States 1992." (1993) *Morbidity and Mortality Weekly Report,* Vol. 41, No. 55.

"Summary of Notifiable Diseases, United States 1993." (1994) *Morbidity and Mortality Weekly Report,* Vol. 42, No. 53.

"Summary of Notifiable Diseases, United States 1994." (1995) *Morbidity and Mortality Weekly Report,* Vol. 43, No. 53.

"Summary of Notifiable Diseases, United States 1995." (1996) *Morbidity and Mortality Weekly Report,* Vol. 44, Nos. 51, 52.

Sutton, Jr., R. (1956) *Diseases of the Skin.* St. Louis, MO: C. V. Mosby Company.

Telford, III, S. R., F. S. Kantor, Y. Lobet, S. W. Barthold, A. Spielman, R. A. Flavell and E. Fikrig (1995) "Efficacy of human Lyme disease vaccine formulations in a mouse model." *Journal of Infectious Diseases,* Vol. 171, p. 1368–1370.

Tilton, R. (1994) "Laboratory aids for the diagnosis of *Borrelia burgdorferi* infection." *Journal of Spirochetal and Tick-borne Diseases,* Vol. 1, No. 1, p. 18–23.

Tilton, R. and R. W. Ryan. (1991) "Laboratory detection of *Borrelia burgdorferi* infection." *Clinical Microbiology Newsletter,* Vol. 13, p. 9.

Tilton R. C. and R. W. Ryan. (1993) "The laboratory diagnosis of Lyme disease." *Journal of Clinical Immunoassay,* Vol. 16, p. 208–214.

Tsai, T. (1992) "Arboviral diseases of North America." *Textbook of Pediatric Infectious Diseases.* [Feignin, R. and J. Cherry. (Ed.)] Philadelphia, PA: W. B. Saunders, p. 1390–1423.

Tsai, T. (1992) "Arboviral infections: General considerations for prevention, diagnosis, and treatment in travelers." *Seminars in Pediatric Infectious Diseases,* Vol. 3, No. 1, p. 62–69.

Update: Lyme disease vaccine. (1991) Fort Dodge, IA: Fort Dodge Laboratory.

Vanderhoof, I. and K. Vanderhoof-Forschner. (1993) "Lyme disease: The cost to society." *Contingencies,* January/February, p. 42–48.

Wahberg, P., H. Granlund, D. Nyman, J. Panelius and I. Seppälä. (1993) "Late Lyme borreliosis: Epidemiology, diagnosis and clinical features." *Annals of Medicine,* Vol. 25, p. 349–352.

Waldholz, M. (1990) "Lyme disease vaccine is tested succesfully on mice at Yale lab." *Wall Street Journal,* October 26.

Wallis, R. C, S. E. Brown, K. O. Kloter and A. J. Main Jr. (1978) "Erythema chronicum migrans and Lyme arthritis: field study of ticks." *American Journal of Epidemiology,* Vol. 108, p. 322–327.

Weber, K. and W. Burgdorfer. (Ed.) (1993) *Aspects of Lyme Borreliosis.* Heidelberg, Germany: Springer-Verlag. Chapters: Åsbrink, E., A. Hovmark and K. Weber. "Acrodermatitis chronica atrophicans." p 193–204; Burgdorfer, W. "The historical road to the discovery of *Borrelia burgdorferi.*" p. 21–28; Herzer, P. "Joint manifestations." p. 168–184; Herzer, P. "Therapy of joint manifestations." p. 340–343; Hovmark, A., E. Åsbrink, K. Weber and P. Kaudewitz. "Borrelial lymphocytoma." p. 122–130; Mayer-Berger, W., M. R. van der Linde and D. Hassler. "Therapy of Lyme carditis." p. 344–349; Pfister, H.-W., W. Kristoferitsch and C. Meier. "Early neurological involvement (Bannwarth's syndrome)." p. 152–167; Pfister, H.-W., W. Kristoferitsch and B. Sköldenberg. "Therapy of Lyme neuroborreliosis." p. 328–339; Preac-Mursic, V. "Antibiotic susceptibility of

Borrelia burgdorferi in vitro and in vivo." p. 301–311; Preac-Mursic, V. and B. Wilske. "Biology of *Borrelia burgdorferi.*"; Schönherr, U. and F. Strle. "Ocular manifestations." p. 248–258; van der Linde, M. R. and P. E. Ballmer. "Lyme carditis." p. 131–151; Weber, K. "Therapy of cutaneous manifestations." p. 312–327; Weber, K. and W. Burgdorfer. "Therapy of tick-bite." p. 350–351; Weber, K. and H.-W. Pfister. "History of Lyme borreliosis in Europe." p. 1–20; Weber, K., H.-W. Pfister and C. D. Reimers. "Clinical overview." p. 93–104; Weber, K., U. Neubert and S. A. Büchner. "Erythema migrans and early signs and symptoms." p. 105–121.

Weber, K. and W. Burgdorfer. (1994) "The cradle of Lyme borreliosis." *Journal of Spirochetal and Tick-borne Diseases,* Vol. 1, No. 2, p. 35–36.

Weber, K., G. Schierz, B. Wilske, V. Preac-Mursic, W. Burgdorfer and A. Barbour. (1983/84) "Antibodies against *Ixodes dammini* and *Ixodes ricinus* spirochetes in tick-borne disorders." 11th ADF meeting, Kiel, Nov 1983. *Archives of Dermatological Research,* (1984) Vol. 276, p. 260.

Whitmire, W. and C. Garon. (1994) "Induction of B-cell mitogenesis by outer surface protein C of *Borrelia burgdorferi.*" *Journal of Spirochetal and Tick-borne Diseases,* Vol. 1, No. 3, p. 64–67.

Wierenga, D. and J. Barry, III. (1995) "The drug development and approval process." *New Drug Approvals in 1994,* Washington, DC: Pharmaceutical Manufacturers Association.

Wilson, M. H. (1986) "Reduced abundance of adult *Ixodes dammini* (Acari:Ixodidae) following destruction of vegetation." *Journal of Economic Entomology,* Vol. 79, p. 693–696.

Wolfe, D., C. Fries, K. Reynolds and L. Hathcock. (1994) "The epidemiology of Lyme disease in Delaware." *Delaware Medical Journal,* Vol. 66, No. 11, p. 603–613.

Zhang, Z. F. and K. L. Wan. (1996) "Concurrent *Borrelia burgdorferi* and *Babesia microti* infection in humans." *VII International Congress on Lyme Borreliosis.* San Francisco, CA: University of California, Berkeley, p. 164.

AFTERWORD

A Special Message from The Honorable Senator Joseph I. Lieberman

Karen Vanderhoof-Forschner is a constituent of mine, but we did not first meet in our home state of Connecticut. In 1989, shortly after I was sworn in as a senator, I happened to be having lunch in the Senate Dining Room when Karen was introduced to me. After a brief conversation, I asked her to let my office know if there was anything we could do for her while she was in Washington. "As a matter of fact, there is . . . ," she said. And so began a long and fruitful friendship . . . and partnership in the battle against Lyme disease.

Because of her own family's devastating encounter with this disease, Karen was inspired to help make the public, the government, and the medical community itself more aware of Lyme disease—so that citizens could take steps to avoid it, doctors could know how to recognize and treat it, and government and researchers could find new ways to combat it.

Karen Vanderhoof-Forschner is the perfect person to bring us *Everything You Need to Know about Lyme Disease*. Written by a parent, the book comes at a time when so many parents have concerns about what to do when they spot a tick on their child's body. We've all heard something about Lyme disease—thanks in large part to Karen's remarkable public education efforts throughout this decade—but we're still largely in the dark about the practical steps we should all take to protect our families.

That's where this book comes in. You've just come home from a hike in the woods with your kids, and while undressing, you see a tick attached to the area behind a knee or underneath a sock. You look at the tick ruefully, with a foggy awareness of Lyme disease, but no notion of what to do next. Do you burn it off or use nail polish or tweezers? How quickly should you act? What are the symptoms of Lyme disease? Should you call the doctor?

This book answers those questions and does so in what I would call "plain language." It includes everything from the names and addresses of companies that make special tweezers that remove ticks, to government public health agencies around the world that are in the forefront of the fight against Lyme disease. Along the way those interested can learn a lot about the history of the disease, medical treatment, and preventive measures that can protect you and your children.

Ticks are commonplace, and when I was young they were considered little more than an itchy nuisance. Yet, today infected ticks harbor bacteria in their guts that can be injected into the victim. Once there, the bacteria cause a wide variety of symptoms developing over time—from flu-like symptoms and a rash to severe headaches, arthritis, and even cardiac abnormalities.

Recent research has indicated that quick action to remove the tick can ward off much of the danger. But diagnosing the disease and treating infection once it occurs remain difficult. We have made great strides in our understanding over the last 10 years. The Lyme Disease Foundation, created by Karen and her husband, Tom, has made a great contribution to that advance. This book is a valuable contribution that I believe will be useful to parents and health care professionals alike.

Lyme disease is a real, national health threat. From 1982 to 1994 over seventy thousand cases of Lyme disease in forty-eight states were reported to the Centers for Disease Control. In 1994 there was a 58 percent increase in the number of cases over 1993. A study conducted by the Lyme Disease Foundation and the Society of Actuaries determined that Lyme disease could be costing this country as much as $1 billion a year in health care costs and lost productivity.

I have worked with Karen to improve awareness of the disease by sponsoring Lyme Disease Awareness Week each year. And we've lobbied, with some success, for funding of Lyme disease research and education. But the Centers for Disease Control estimate that thousands of Lyme disease cases still go undiagnosed, unreported, and untreated. That's why efforts by individuals and private organizations like the Lyme Disease Foundation are critical. And that's why *Everything You Need to Know about Lyme Disease* is so timely.

I wish this book could be supplied as a gift to every parent leaving a hospital with a newborn child—or, better yet, to every man and woman for their library shelf. But for now, read and learn, and buy a copy for someone you love . . . because there may well come a day when they'll be happy to have it on their shelf.

—Senator Joseph I. Lieberman
Connecticut

INDEX